The Cat's Meow

The Cat's Meow

How Cats Evolved from the Savanna to Your Sofa

Jonathan B. Losos

Illustrations by David J. Tuss

VIKING

VIKING
An imprint of Penguin Random House LLC
penguinrandomhouse.com

LIBRARY OF CONGRESS CATALOGING-IN-PUBLICATION DATA
Names: Losos, Jonathan B., author.
Title: The cat's meow : how cats evolved from the Savanna to your sofa / Jonathan Losos.
Description: First edition. | New York : Viking, an imprint of Penguin Random House LLC, [2023] | Includes bibliographical references and index.
Identifiers: LCCN 2022034185 (print) | LCCN 2022034186 (ebook) | ISBN 9781984878700 (hardcover) | ISBN 9781984878717 (ebook)
Subjects: LCSH: Cats. | Cats—Behavior. | Cats—History.
Classification: LCC SF442 .L645 2023 (print) | LCC SF442 (ebook) | DDC 636.8—dc23/eng/20220822
LC record available at https://lccn.loc.gov/2022034185
LC ebook record available at https://lccn.loc.gov/2022034186

Printed in the United States of America
1st Printing

DESIGNED BY MEIGHAN CAVANAUGH

To my late father, Joseph Losos,

to whom I owe my love of cats

and so much more

CONTENTS

The Cat's Meow

One

The Paradox of the Modern Cat

It's a good thing cats aren't the size of large dogs, says an old joke, because if they were, they'd eat their owners. As a cat-loving scientist, my first reaction was to laugh, quickly followed by the thought "How can I research this idea?" Sadly, even science has its limits. Until we're able to produce seventy-five-pound housecats, we'll never get a definitive answer.

That's not to say that science is entirely silent on the question. A 2014 research paper was widely reported to conclude that "Cats Would Kill You If They Were Bigger," as the *Orlando Sentinel* declared in its headline. *USA Today* eliminated the subtleties, declaring "Your Cat May Want to Kill You."

In reality, the research paper said no such thing. Rather, the scientists compared behavioral tendencies, such as aggressiveness and sociability, among five feline species ranging in size from the housecat to the African lion. The paper's primary conclusion was that personality-wise, there aren't many differences among cats, regardless of size. Zookeepers have told me the same thing: if you can read the expressions and body posture of your cat, you can understand what a lion or tiger is thinking. The researchers did not suggest that housecats, were they the size of lions, would be sizing you up for supper—it was the journalists and bloggers who made that leap.*†

Regardless of its implications for potential man-eating moggies,‡ the research reveals an important fact: in many ways, a cat's a cat, whatever its size. This finding will come as no surprise to anyone who's spent hours watching internet videos of tigers chasing laser-pointer dots, leopards jumping into cardboard boxes, and lions rolling in catnip.

The point that our household friends are little different from their wild relatives was driven home to me a few years ago on a trip to South Africa with my wife, Melissa. While driving around at night near Kruger National Park, a common sighting was a slender, tawny feline, faintly spotted or striped, the small cat caught in the spotlight's glare for just a moment before he darted back into the shadows.

The first few I saw were reasonably close to the game lodge where we

---

*On the topic of eating you, a number of these journalists also pointed out that when people die in their homes and are not discovered, dogs eat their owners' bodies much more frequently than cats. One medical journal, for example, included gruesome case studies of three corpses eaten by dogs but only one by cats.

†A footnote on footnotes: my head is full of all kinds of fascinating information and insights on cats. I have relegated points that are tangential or provide extra details to the footnotes. You are free to ignore them, but beware: you're going to miss some great stuff. References and some commentary are available in the Notes on Sources at the end of the book; a much more extensive list of references can be found at www.jonathanlosos.com/books/the-cats-meow-extended-endnotes.

‡For non-Brits, "moggy" is a word used in England for "cat," often specifically a non-pedigreed cat.

were staying. Based on their size and appearance, I presumed they were pets that belonged to one of the staff members, or perhaps the lodge kept them to keep rodents in check. In any case, they certainly seemed to be domestic cats, out for a ramble in the African wilderness. No good can come of this, I thought, with all the larger predators about, but that's their business, not mine. So I didn't pay much attention to these small waifs, nor was I disappointed when they quickly disappeared back into the bush—I'd try to give them a nice pat when I saw them back in camp.

An African wildcat.

But one day we encountered one of these cats miles from the lodge, and I realized that this could not be anyone's pet. And, indeed, he wasn't—he was an African wildcat, the species from which domestic cats arose (we'll discuss in chapter six how we know this).* Further scrutiny revealed distinctive features: legs longer than most domestic

*We'll also discuss recent research that shows that what was once known as the African wildcat is actually composed of two genetically different forms, the North African and South African wildcats. Because the two are very similar and older writings didn't distinguish between them, I'll often simply refer to the "African wildcat."

cats and a striking black-tipped tail. Still, if you saw one from your kitchen window, your first thought would be "Look at that beautiful cat in the backyard," not "How'd that African wildcat get to New Jersey?"

In terms of behavior, too, most domestic cats differ little from their ancestors. Sure, they're friendlier—or at least more tolerant—of humans, and sometimes more sociable to each other, but in other ways—their hunting, grooming, sleeping, and general manner—they behave just like wildcats. Indeed, the ease with which abandoned cats go feral and revert to their ingrained, ancestral ways is evidence of how little the domestic cat has evolved.

For this reason, domestic cats are commonly referred to as "barely" or "semi-domesticated." Domestication is the process by which animals and plants are modified by their interactions with humans in a way that benefits us.* By "modified," I mean they have evolved through genetic changes that result in behavioral, physiological, and anatomical differences from their ancestors.†

In contrast to cats, "fully domesticated" species are substantially different from their wild ancestors. Consider the barnyard pig. Big, portly, pink, curly tail, floppy ears, very little hair. *Sus domestica* is the quintessential domesticated animal, a species sculpted by humans, greatly modified from the ancestral boar (*Sus scrofa*) to suit our needs and desires. Or contemplate cows, far removed from their majestic wild cattle ancestors, turned into meat- and milk-producing machines

---

*Like so much in academia, scholars argue vehemently over what constitutes domestication, whether domesticated species benefit in some ways from being domesticated, and other related topics. Check the endnotes for references that provide an entrée to this literature.

†Evolution in a biological sense is a genetically based change in a population through time. Such changes during domestication are just as much evolution as what goes on in nature.

by our selective breeding over the millennia.* Similar selection applied to plants has created food crops like corn and wheat that are vastly different from their wild progenitors.

Not so for domestic cats. Look underneath the paint job—the variation in hair length, color, and texture—and most domestic cats are nearly indistinguishable from wildcats. The many great differences in anatomy, physiology, and behavior that distinguish most domesticated species from their ancestors don't exist in cats.

Recent genome studies confirm this view. Whereas dogs have diverged from wolves in many genes, domestic cats and wildcats differ in only a handful. Cats truly are scarcely domesticated.

But this statement comes with a major asterisk. A small minority of cats are members of specific breeds (the rest are lumped into the category "domestic shorthairs and longhairs," a more polite equivalent of "mutt").† A breed is a group of individuals that share a distinctive set of traits that distinguish them from other members of the species. The distinctiveness of a breed is maintained by mating members of a breed only with each other generation after generation, firmly establishing the genes for these traits throughout the breed.‡

Cat breeds vary in how distinctive they are. Some deviate little from the standard model, looking like a typical domestic cat, perhaps differing slightly due to curly hair or floppy ears.

---

*For example, modern cows have enormous udders, capable of producing up to eight gallons of milk a day; in contrast, the mammae of the cow's ancestor, the auroch, were barely noticeable.

†"Randombred" is another term for these cats.

‡More technically, a "breed" is defined as "a stock of animals or plants within a species having a distinctive appearance and typically having been developed and maintained by deliberate selection by humans." We'll see in chapters thirteen and fourteen that breed development and maintenance in some cases can be more complicated than I've just described.

But many cat breeds diverge greatly from the ancestral physique and behavior. If you happened upon a member of one of these breeds on the African savanna, you'd never mistake her for an African wildcat.

Indeed, some breeds are remarkably different not only from a standard housecat, but also from all other members of the Felidae (the scientific name for the cat family, encompassing everything from domestic cats to ocelots, lions, and tigers). Selective breeding, in other words, has created cats unlike those produced by millions of years of feline evolution.

So we have a cat conundrum. Most pusses are little changed from the ancestral version, yet a minority are radically different. How can cat evolution be simultaneously in slow and fast gear? Clearly, *Felis catus*—the domestic cat—is not a monolithic entity, evolving as one. Quite the contrary, multiple realms of cats exist, and these realms are evolving in very different ways.

To understand why, we have to think about the different categories of cat living around us. On the one hand, household pets are divided between those that are members of specific breeds and those that aren't. On the other hand, unowned cats, those that don't live in people's homes, can also be split into two groups, those living entirely on their own and those being fed and taken care of by people (at least to some extent).*

The possibility that different groups of cats are evolving in different ways opens the door to questions about the future. Now that cats have traded the savanna for human environs, are we witnessing the origin of species, the domestic cat splitting into multiple lines, each going its own evolutionary way?

---

*Some people more finely divide unowned cats into many categories, but the dichotomy I present captures the key differences for our purposes.

To address these questions, let's consider the types of selection acting on these groups, starting with the cats belonging to specific breeds. Charles Darwin recognized that the way animal and plant breeders work is analogous to what goes on in nature: individuals with certain traits survive and reproduce more than individuals without those traits. If the traits are genetically based—that is, if individuals with the traits have different gene versions than individuals without them—then those versions of the genes and the traits they produce will become more common in the next generation. Continued over many generations, such selection can lead to substantial change. That is the way that species evolve by natural selection in nature, and it is the same process—called artificial selection at the hands of humans—underlying the development and refinement of new breeds.*

We'll postpone for now discussion of why breeders choose to select for the specific traits they favor and why they even bother to create new breeds in the first place. The salient point is that breed development is an evolutionary process that produces plants and animals with entirely new features or new combinations of existing features. Because all members of a breed possess the genes responsible for these features, a breed retains its distinctiveness from one generation to the next. That's why breeders talk about the "pedigree" of a specific individual—it demonstrates that an individual is descended from multiple generations of ancestors who are also members of that breed, and thus the individual should reliably possess the breed's particular features (for this reason, I will refer to cats that are members of a breed as pedigreed).

---

*In fact, much of Darwin's evidence for the effectiveness of natural selection in *On the Origin of Species* came from examples of the breeding practices of farmers and hobbyists (not surprisingly, since no one was studying evolution in nature at the time). Darwin subsequently wrote an entire book on the subject, *The Variation of Animals and Plants Under Domestication*.

Most owned cats, however, are not members of any breed (about eighty-five percent of pet cats in the United States, compared to fifty percent or fewer of all dogs). These are the cats in most people's homes and that you see in pet stores and animal shelters. They may be a genetic mix of several breeds or, much more commonly, cats with no pedigreed ancestry at all. As a group, they have no defining characteristics other than being domestic cats. If you tell me you have a domestic shorthair, the only thing I know about him is that he's a cat with short hair. By contrast, tell me you have a member of a specific breed—say, a Singapura—and I immediately have a picture in my head of what your cat looks like and maybe even how she behaves.

Most importantly for our purposes, the majority—more than ninety percent—of household pets in the United States are neutered and so aren't contributing their genes to the next generation. They're an evolutionary dead end. The pet cats we have in our houses are the result of the evolutionary descent from the African wildcat, but most of them are not shaping the future evolution of the species.

Rather, most reproduction among non-pedigreed cats is going on outside the house, in alleys, woods, and farms, out of our control. The cats themselves are deciding who breeds and who doesn't. No artificial selection there. We are not choosing which ones can breed and which ones can't, so there is no selection for traits that we might favor.

Some of these cats exist on their own, away from and independent of humans. They live lives very similar to the ancestral wildcats, and we'd expect natural selection to be favoring them to be just the way they are, hewing to the formula that has made wildcats successful for millions of years.*

Many unowned cats, however, live in proximity to people, often in-

*Selection to remain the same is called "stabilizing selection."

teracting with and being fed by us. For these cats, we can imagine a mix of selective pressures, in some ways favoring the ancestral wildcat ways for survival outside, but also favoring traits beneficial for living around and mooching off us.

Of course, even better than imagining would be scientific data on how natural selection is shaping these cats. As we'll discuss later on, surprisingly little research has been conducted on how natural selection affects these cats, but the time is ripe for that to change.

This, then, is the sense in which the modern cat world is divided, one prong entering new evolutionary territory, producing felines unlike any the world has ever seen, perhaps even qualifying as fully domesticated. Meanwhile, many cats live a lifestyle not very different from their ancestors, out in nature, dealing with the elements, interacting with other species, impacting ecosystems. Because they are in charge of their love lives, they are determining their own evolutionary future, and not surprisingly, they're sticking with the time-tested wildcat blueprint. An intermediate third prong is splitting the difference—unowned cats adapting to deal with the rigors of outdoor living, but also exploiting us as a food-providing resource.

This book is about how cats have come to this evolutionary fork in the road. It explores how selection—both natural and artificial—over the last several millennia has shaped the contemporary cat and continues to do so today; how cats, in turn, are interacting with the world around them; and what the future may hold for *Felis catus*.

THAT'S ALL WELL AND GOOD, but how do I, an evolutionary biologist who specializes in studying how lizards adapt to their environment, come to be writing this book? I'll confess that I've always loved cats, ever since as a five-year-old I went with my mother to the Animal Protective Association of Missouri and adopted a Siamese cat to surprise my father on his birthday. I still remember when he came home from work, trying to hide Tammy behind my thin little-boy legs as dad walked into the kitchen, but her meowing gave away the surprise. Ever since, I've been gaga about felines.

But as I set out to become an evolutionary biologist, it never occurred to me to study cats. Their famed secretiveness was not attractive to someone who wanted to get out into nature and observe animals as they go about their daily lives. Lizards seemed so much more manageable: abundant, easy to find, easy to work with in both the field and the lab. I went with lizards and didn't look back.

As I set about building my career, I didn't pay much scholarly attention to cats, although I did pet them whenever possible. My impression was that there wasn't a lot of cat research going on, and the little that existed wasn't very interesting.

Turns out I was wrong. Several years ago, I learned that researchers were studying domestic cats employing all the approaches that I and my colleagues use to study lizards, lions, elephants, and other wild species. Everything from kitty cameras to GPS tracking to genome sequencing. I was surprised and amazed. Who knew there were so many scientists interested in cats, much less how much they had learned about the biology of our little friends?

And then I had what I humbly submit was a great idea. I'd teach a freshman college course on the science of cats. The goal would be to attract the students with the topic of cats, and then, when they weren't looking, I'd teach them a lot of cutting-edge science—ecology, evolution, genetics—while they thought they were learning about felines.

It worked like a charm. Twelve fabulous Harvard freshmen enrolled in my class. We had a guest lecture by an Egyptologist on ancient cats, visited a cat show on Cape Cod, inspected portraits of cats at the Fogg Art Museum, and fed stray cats at the break of dawn behind boarded-up houses in south Boston. We learned a tremendous amount about cats, of course, and along the way, the students learned how modern-day biologists study biodiversity.

But something unexpected happened as well. Even as I was using cats to teach the students about science, I was myself becoming enamored with cat science.

I'm particularly fascinated by the diversity of modern-day cat breeds. Much of my research on lizards has focused on how, over thousands and millions of years, a single ancestral species can give rise to a large number of descendant species, each anatomically and behaviorally specialized to use a different part of the environment (technically, the phenomenon is called "adaptive radiation"). In comparison, the diversity of cats, which has occurred in decades rather than millennia, is extraordinary.

The November 1938 issue of *National Geographic* featured an article on cats, with photographs of Persian and Siamese cats that didn't look all that different from each other. You certainly couldn't say that about those breeds today. Just eighty-five years later, Siamese have gone from normal-looking cats with somewhat angular heads to extraordinarily elongated, slender, "slinky" cats with heads the shape of a spear point. It's as if someone grabbed a 1938 Siamese and stretched his nose far forward from his eyes. Persians have been altered in the opposite direction to produce short-bodied, heavyset cats that basically don't have a nose at all. In other words, in just a few decades, breeders have reshaped the anatomy of these felines to produce cats extraordinarily different from each other and unlike any cats that have ever lived. Or consider the short-legged Munchkin breed. If paleontologists found a fossil cat with that anatomy, they would likely classify it as something very different from *Felis catus*.

Cats, in other words, are a great example of evolutionary diversification, one that has occurred extremely rapidly and is worthy of scientific study. Realizing that, I haven't given up my day job as a lizard researcher, but I am now also studying cats, investigating what we can learn about how they've evolved and continue to evolve, and what that might tell us about the evolutionary process in general.

CAT SCIENTISTS HAVE a severe case of dog envy. And for good reason: dogs have become the darlings of laboratory scientists and the journalists who report on them. If *The New York Times* is to be believed, canine studies are at the cutting edge of modern science, while feline investigation is stuck in the Middle Ages. Indeed, research on dogs has yielded important advances in some areas, such as genetics. Cat studies, though with less media coverage, have been equally bountiful, not only in many of the same areas as dog work, but also on topics untouched by canine scholarship. In many ways, it's a golden age for the scientific understanding of our beloved pets. Thanks to the marvels of modern technology, the many mysteries of the cat are rapidly yielding to a new generation of ailurological* investigators.

The results of this research provide the material for this book. To understand today's cats, we need to know their roots, who their ancestors were, how they've changed, and why. Archaeology, genetics, behavioral observations, and audiospectral analysis are just some of

*The English language is sorely lacking a word for the study of cats. Some online sources call it felinology, but if you're going to make up a word, you might as well get it right ("felinology" mixes Greek and Latin). "Ailurophile" and "ailurophobia" are accepted dictionary words, meaning the "love" and "fear" of cats, respectively, and are derived from the ancient Greek αἴλουρος [*aílouros*] for "cat" (literally, it means "with waving tail"). It follows, then, that the study of cats should be termed "ailurology."

the types of research that can inform us about how cats have evolved over the last ten thousand years. We'll also explore the high-tech approaches scientists are using to study how today's cats interact with their environment—what they do when they go out the back door and head off to whereabouts unknown. The health of the environment is part and parcel of such discussions, and we'll touch on the impact cats have on other species and what we can do about it. And finally we'll wrap up by discussing the cat of the future: where felinedom is headed and what possibilities lie ahead.

The Cat's Meow will also explore what we don't know about cats. Certainly, there is no shortage of information available in the myriad of books, websites, and magazines devoted to all matters moggy. But as a scientist, I find myself often frustrated; it's hard to separate fact from urban legend. Consider, for example, the names of many cat breeds. Is the Egyptian Mau really a little-changed descendant of the cats of the pharaohs? And do Persians, Abyssinians, Siamese, and Balinese really hail from their nominal places of origin?*

I often wonder, too, about the evolutionary explanations put forward by cat whisperers for the wacky things cats do. Does Bella really leave dead mice on your pillow to help you improve your hunting skills, and do cats truly chatter at the window because the sight of birds triggers the same rapid jaw movements they use to dispatch their prey? And what about Mr. Whiskers making biscuits on your belly? It's equal parts adorable and annoying, but why does he do it? Concocting evolutionary just-so stories to explain why a particular species possesses a particular trait is easy. But scientifically testing such ideas is often much more difficult. For this reason, we'll need to consider not only what we know about the cat's evolutionary journey, but

*Spoiler alert: the answers are unlikely and sort of, it's complicated, yes, and no.

also what we have yet to discover and what questions science may not be able to answer at all.

Of course, the saga of cat evolution involves people as well as well as felines. As we'll see, for thousands of years, our role was unintentional; cats drove the bus, evolving on their own terms to live around us. But in the last fifteen decades, the shoe has been on the other paw, and in many ways we have pushed the evolution of cats—at least some of them—in very new directions. We'll explore how and why cat fanciers desire to create new breeds and what science tells us about how it happens.

The breeding and purchase of pedigreed cats comes in for a lot of criticism in some quarters. We'll examine these critiques, some of which have merit. At the same time, we'll consider whether selective breeding holds the prospect of producing cats that are better suited to living as domesticated creatures in today's modern world.

The cats, of course, will be the central focus of *The Cat's Meow*. To understand them, we'll go to where they live and where they're being studied, from suburban bedrooms to scientific laboratories, resort islands, and the Australian Outback. While there, we'll meet the people involved in the cat world, the scientists and breeders who for one reason or another ended up spending their lives with a feline focus.

People have strong feelings about cats, so deciding how to refer to them can be tricky. There's not much controversy in referring to a cat as a "pet." But what do we call the human side of the relationship? As the saying goes, "Dogs have owners, cats have staff." In a less jocular way, many consider their feline companions to be more akin to friends or family than possessions. "Cat owner" has been discarded in many circles.

An increasingly common alternative is "pet parent." Although I understand the reason some like this phrase,* I will not use it because I want to emphasize that cats are neither mini-lions nor mini-people. They're cats!

"Friend," "companion," and many other names are also used. To my mind, there's some truth in all of these, and none of them is perfect. I'll use various terms interchangeably.

What's more important to me is emphasizing that cats are not objects. They're living, sentient beings, and we certainly interact with them as such. For this reason, I will not use "it" in reference to a cat. To avoid using "he/she," or similarly awkward conventions, I'll alternate between referring to generic cats as females in even-numbered chapters and males in odd-numbered chapters. Of course, when discussing specific, real cats, I will refer to them according to their real sex.

There's also the question of what name we use for the species. Scientifically, *Felis catus* does the trick, but what about in everyday usage? I grew up calling them housecats. Some people, however, get their knickers in a twist about this term. After writing a piece for *National Geographic*'s website in which "housecat" was used, I received a haughty letter informing me that the term only applies to cats that never go outside.† The writer of this letter—and his pedantic fellow travelers—are welcome to their opinion, but in fact, the term is commonly applied to all members of *Felis catus*, regardless of abode. On the other hand, "domestic cat" is also widely used, but to my mind is both dull and sometimes misleading: there's very little "domestic" about many members of this species.

---

*Whereas others hate it—do some googling and check out the nasty online debate.

†The website used "house cat" in the title, but "housecat" in the text, illustrating that the term has two, widely used, spellings. I prefer "housecat" referring to the entire species, leaving "house cat" to refer to a cat that lives solely in a house.

We could, and often do, refer to them simply as "cats." But we also use that term to refer to all members of the cat family, from lions to lynxes (why dogs got their own name, separate from other members of the Canidae, and cats didn't is an interesting question I won't pursue).

My solution? I'll use all three interchangeably. I'll tend to use "cat" when it's clear to whom I'm referring, but when there's possible ambiguity as to which feline, I'll use "housecat" or "domestic cat" as the whim hits me.

Finally, a huge variety of terms have been created to refer to cats living in different ways. Although there are subtle distinctions responsible for the profusion of categories, I'm simply going to refer to "pets" and "unowned cats," recognizing that there is a gray zone between them. Among "unowned cats," those that live in large groups and are fed by people may be referred to as "colony cats." Those that live on their own, usually by themselves and not being fed or otherwise taken care of by people, will be referred to as "feral cats," again recognizing that the line between feral and colony cats can be blurry. A common distinction between "stray" and "feral" cats is that strays are socialized to humans due to previous interactions, whereas feral cats are not and react fearfully to our presence. Given enough time on their own, stray cats can lose their socialization and become feral.

ENOUGH TERMINOLOGY! Most books on cats open in ancient Egypt, talking about how the African wildcat came to live among humans and then was domesticated, first as a mouse catcher, then as a pet, finally as a god. It's a great story, and we'll get to it eventually. I want to start off, however, on a different tack, by discussing the cats of today.

I've already discussed how little most cats have changed from their ancestors, but that doesn't mean they haven't evolved at all. We'll start by considering the few ways non-pedigreed cats have changed, the "semi" in their domestication. For many species, the first step in the domestication process is change in comportment and temperament, and so that's where we'll begin, by considering the behavioral changes *Felis catus* evolved in its descent from the African wildcat.

Two

The Cat's Meow

Meowww!

That's a sound familiar to anyone who's ever lived with a cat, and to most people who haven't. The meow is the quintessential cat trait, the defining characteristic of catdom to many. But just what are cats trying to say when they meow, and to whom? And if they're talking to us, does that mean the meow is a housecat trait that evolved during domestication?

I ALWAYS ASSUMED that cats communicate with each other by meowing, and that they've just expanded their social circle to include us. The cat experts, however, say otherwise. Look up any scientific review of cat communication and you'll read that when adult cats interact with each other, they rarely meow (though they will make other sounds, especially in unfriendly interactions).

But there's a funny thing about these scientific papers. When scientists make statements in a paper, they reference other papers that have provided evidence for that statement (except when interpreting their own data presented in the paper). And if you look at all the papers that say that cats don't meow to each other very much, you'll find that they all refer to a single study.

It would seem, then, that the entire evidence that cats rarely meow to each other comes from one study of colonies of neutered cats living outdoors in England. I have no reason to question this finding—I can't recall any of my cats ever meowing to each other (as opposed to the occasional hiss or growl).

Still, I wondered how safe it was to extrapolate from one study of outdoor cat colonies to all cats that share our homes. Taking that idea one step further, I thought about what sort of cat, if any, would talk to her fellow felines. And then it came to me: surely it would be Siamese cats, the chatterboxes of the feline world, renowned for talking to humans nonstop. But do they do the same to each other?

Lacking both the time and funding to conduct a proper scientific investigation, I turned to the next best thing: an informal survey on Facebook. To the private group Siamese Cats, I posted: "Siamese are famous for talking with their humans. But do your cats ever meow to each other?"

The results were lopsided: twenty-eight of forty-one respondents (sixty-eight percent) said their Siamese meowed to other cats. A number of the "no" respondents noted that their Siamese made other sounds—like chirps—to other cats but didn't meow.

Could all the Siamese aficionados who responded positively be mistaking a different type of vocalization for a meow? Or could these cat cohabitators—never trained to study animal behavior—be mistaken when they conclude that Jasmine's meow was directed to Sheba?

Maybe. But on the other hand, people spend a lot of time around their pets and know them well. When two thirds of respondents say their cats meow to each other, this possibility needs to be taken seriously.

Two observations bolster the conclusion that Siamese actually are talking to each other. Ten of the twenty-eight "yes" respondents specifically said that when their cat couldn't locate other cats in the household, she started meowing to try to find them. Also, several "yes" respondents said that when their Siamese meowed loudly, others came running to her.

Tangentially, there was one other funny point from the survey: several respondents said their Siamese meowed to their dog!

I ran these results by several experts on cat social behavior and communication who agreed that these results rang true.* Although cats do meow much more to us, they also may meow to each other (and to dogs) at times, especially when trying to find each other.

Regardless, the question of interspecies communication remains. Cats meow more to us than they do to each other, so it's not just a case of them treating us as part of the clan. What are they trying to tell us?

ANYONE WHO'S LIVED WITH A CAT knows that there's no "one-size-fits-all" meow—rather, a cat has a diverse meow-cabulary with different sounds used in different situations. Other animals have sim-

*To be fair, though, the experts also emphasized that people may have been mistaken in interpreting what they observed.

ilar vocal repertoires: dogs vary their barks in different situations, monkeys have different alarm calls for different predators.

If I were a betting man—and I am!—I'd put down good money that the different meows of cats are meaningful. And that's just what a graduate student at Cornell University did.

Nicholas Nicastro enrolled in graduate school to study the evolution of human language. Finding the anthropology department riven by academic infighting and political correctness, he transferred to the psychology department, planning to work on primates.

One afternoon, he was conversing with his doctoral advisor, an expert on animal communication. The professor proclaimed that animal calls may have some emotional content, but nothing resembling a language. Nicastro countered that he had grown up around cats and thought that different meows were meaningful. His advisor was dubious and a wager was formulated. The idea of doing a PhD on primates was out—cat communication it would be.

Nicastro recorded meows from his own two cats and those that lived with a dozen friends and relatives. To make the recordings, Nicastro went to the cats' houses and hung around until the cat was accustomed to his presence, which usually took an hour. Using a microphone placed within six feet of the cat, he then recorded meows made by the cat when she was being friendly toward her human housemate, was about to be fed, and was being brushed vigorously. Recordings were also made when the cat was placed behind a door or window she wanted to go through and when she was placed in an unfamiliar environment (specifically, inside Nicastro's car).

Sometimes it became quite tedious waiting for an uncooperative cat to meow. In addition, when the cats were brushed, Nicastro ran his hand backward against the lay of the hair, trying to induce a meow indicating aggravation or unhappiness. Instead, several times he was bitten. Otherwise, the work went smoothly and Nicastro recorded more than five hundred meows.

The question he was investigating was simple: could people listen to a call and correctly identify the situation in which it was made? Nineteen college students participated for course credit; another nine young adults received a chocolate treat.

Participants came to the testing room, put on headphones, and listened to the cat calls one at a time, pushing a button on a console to indicate their guess at the context in which each meow was uttered.

Overall, the participants revealed only a slight ability to successfully identify the context: twenty-seven percent of calls were correctly identified, whereas by sheer luck, given that there were five possibilities, one would expect twenty percent of the answers to be right. Participants who had more experience living with cats, or who indicated they really liked cats, did better, but even the most successful participant, a woman who lived with cats, loved them, and interacted with them frequently, only correctly identified the context forty-one percent of the time. Nicastro lost his bet on the meaningfulness of the meow, but ended up with a great dissertation.

Similar results were produced by a subsequent study in Italy in which participants heard meows produced by cats that either were waiting to be fed, had been introduced into a room in an unfamiliar house, or were being nicely brushed by their human companion. Overall, the accuracy of the respondents was no better than guessing, although women were somewhat better than men, and people who had lived with a cat were more accurate than those without the experience of cat cohabitation.

These findings suggest that people have only a limited ability to distinguish the contextual meaning of a cat's meow. The results are puzzling because we know that any given cat has a large repertoire of different meows. Moreover, in similar studies with dogs barking in different situations, people have a much greater ability to identify the context. Why can't people figure out what a cat's trying to say?

The answer was revealed in a study conducted in England in 2015. Using an approach similar to Nicastro's, the researchers went to people's homes and recorded cats meowing in four different contexts.* Then they played the calls back to listeners to see if they could correctly identify the context of each call. An important difference from Nicastro's study, however, was that people who lived with each cat were included among the listeners.

Participants were reasonably proficient when listening to the cat with whom they shared a home, correctly identifying the context sixty percent of the time. By contrast, when hearing an unfamiliar cat, they picked the correct context a paltry twenty-five percent of the time, no better than guessing randomly.

These results suggest that each cat has her own specific meows that she uses in different situations, and that people who live with these cats learn to recognize what each meow means. However, these calls are cat-specific; there is no universal cat language, with one type of meow proclaiming "I'm hungry" and another indicating "I'm scared."

How these differences arise is unknown. Some scientists speculate that cats try out a variety of calls and learn which ones get the best response from their human companions in particular circumstances. Plausible, but I'm unaware of any data to back that up. Whatever the reason, the result is a private vocabulary shared between each cat and her human housemates.

---

*Amusingly, the university's ethics committee would not allow the researchers to brush the cats backward, presumably because it was considered too mean or annoying.

Keep in mind that these points pertain only to the sound we call a meow. Cats make many other sounds, and the meaning of some of them—hisses and growls, for example—are quite clear to anyone. Swedish ailurologist Susanne Schötz, perhaps the world's leading authority on cat vocalizations and winner of the prestigious Ig Nobel Prize (given to "honor achievements that first make people laugh, and then make them think"), lists eight types of call—meow, trill, howl, growl, hiss, snarl, purr, and chirp—as well as combinations among them, such as the growl-howl or trill-meow.

As for what, exactly, constitutes a meow, here's what she has to say: "As a rule, meows are produced with an opening-closing mouth. . . . The *m* is produced with a closed mouth, the mouth then opens for the *e* and stays open for the *o*, before closing with the *w*. Try meowing yourself, and look in the mirror while you do it. Do you see how the mouth first opens and closes?"

She's right—that's how I meow. But Schötz goes on to point out that the meow sound is hypervariable. Sometimes, instead of starting with an *m* sound, cats substitute a *w* or *u*. And sometimes they add extra syllables, like "meow-ow" or "me-o-ow." The range of permutations is almost endless. Schötz provides four subcategories—mew, squeak, moan, and meow—but emphasizes that "a meow can be varied almost without end, and because there are so many different versions, it is not simple to assign the sounds to different subcategories."

The "mew," in particular, deserves discussion. A high-pitched meow in which the last, mouth-closing *w* syllable is omitted, mews are the adorable sound made by kittens when calling to their mothers. Some have suggested that the regular meow that our feline friends so commonly direct toward us is an extension of their juvenile behavior—they mewed to their moms, now they're making a similar, more adult sound to us.

. . .

IF MEOWS ARE USED primarily to communicate with us, what does that say about the origin of meowing? Is it a feature that evolved when cats started associating with humans a few thousand years ago?

Apparently not. Observations on zoo animals indicate that most small feline species also meow.* And like their domestic brethren, these species only very rarely meow at each other, even less commonly than housecats. Another similarity is that kittens of these species mew to their mothers, just like domestic cat kittens.

Unlike domestic cats, however, these wild species rarely meow toward people. In a survey of zookeepers, of the 365 cats cared for, only two—both servals—meowed while near their caretakers. This was not due to lack of friendliness: several other species were amiable in many other ways to their keepers. If these species aren't meowing to each other or to people, with whom are they trying to communicate? Or are they just talking to themselves? This is one of the many mysteries of feline biology requiring investigation.

A particularly important feline to consider, of course, is the African wildcat. By studying the meowing behavior of this species, we can understand which aspects of the housecat's behavior were inherited from her ancestor and which evolved as the housecat adapted to living around humans.

And that's exactly what Nicholas Nicastro set out to discover in the second part of his doctoral research. To do so, Nicastro traveled halfway around the world to the Pretoria zoo, which specializes in breeding African wildcats. A dozen cats were kept in enclosures arranged

*Most large cats, like lions and tigers, are physically unable to meow and instead have structures in their throats allowing them to roar.

side by side, close enough for the cats to interact with each other. Nicastro set up microphones around the cages and observed the cats from the outside.

The project worked beautifully. Thanks to the attentive care of their keepers, the cats were "blasé about humans" and carried on unperturbed by the grad student's watchful presence.

African wildcats definitely meow! "I was surprised that they meow so much. It's pretty much a constant din," he recalls.

In fifty hours of observations, he recorded nearly eight hundred meows, noting the circumstances in which each was uttered. Meows occurred when the wildcats were about to be fed, were engaged in aggressive encounters or were pacing back and forth. Only very rarely did one cat meow when engaging in friendly behavior with a person or another cat. Nicastro only studied adult cats, but other scientific reports indicate that African wildcat kittens mew to their mothers, just like other small felines.

Nicastro wasn't just interested in whether African wildcats meow or not. He wanted to know how those meows compared to those of housecats. To find out, he digitally analyzed the audiospectral qualities of the calls, comparing calls given in similar situations.

Computer analysis clearly indicated the calls of the two species differ in all contexts: housecat meows are higher pitched and shorter in duration. The wildcat calls came across as more urgent, demanding. "Mee-O-O-O-O-W!" as Nicastro put it, compared to the domestic cats more pleasing "MEE-ow."

Nicastro then recruited college students to listen to recordings and say whether they found the calls of domestic or wildcats to be more pleasant. Each student listened to forty-eight calls, twenty-four from each species, and provided a pleasantness score for each on a scale of one to seven. Not surprisingly, they could tell the difference between the Africans and the local cats and overwhelmingly favored the sound of the home team.

Is it a coincidence that to our ears, domestic cats meow in a more melodic way than African wildcats? Nicastro thinks not. He suggests that short, higher-pitched sounds are inherently more pleasing to our auditory system, perhaps because young humans have high-pitched voices, and domestic cats have evolved accordingly to curry our favor.

This is an example of a scientific idea called the "sensory bias hypothesis." The idea is that to communicate effectively, species evolve in ways that correspond to the sensitivities of the receiver's detection capabilities. For example, female frogs are particularly good at hearing sounds of a particular frequency, so males produce mating calls of just that pitch. Similarly, guppies are particularly good at seeing the color orange—perhaps because choice food items are that color—and so flashy males sport orange spots to attract the attention of females. Although these examples refer to interactions within a species, the same holds true for communication across species. In this case, cats may have evolved to take advantage of our preference for high-pitched sounds.

More generally, these data indicate that domestic cats did not invent the meow. Nonetheless, they have adapted the utterance to live with us, altering the meow's sound and using it in different contexts. They are not simply considering us to be fellow cats, communicating with us as they do to their feline brethren, because cats of all species rarely meow when talking to each other. The big difference is that domestic cats have evolved to meow to people as part of friendly interactions and correspondingly have altered the meow so that we find it more appealing.

. . .

THE MEOW ISN'T THE ONLY sound cats make that has been shaped by their association with us. Consider their other marquee sound.

Cats purr in many situations. Not only when they're happy, but when they're waiting for food, stressed, and sometimes even when they're in pain. Just as with their vocalizations, the sound a cat makes when she purrs differs from one situation to the next.

In particular, cats are known to have loud, insistent purrs when they're about to be fed—think about the cat at your feet, perhaps rubbing against your legs, as you open a can of wet food. A team of scientists decided to see what the cats might be trying to tell us when they change their purr.

To do so, they recorded the purrs of ten cats in two situations. The first was when the cat was stroked by her companion at a calm time, producing the contented purr we all cherish. The second was in the morning, feeding time for the cat; but instead of getting up to take care of culinary duties, the cat's human companion was instructed to remain in bed. Up onto the mattress the cat would jump, positioning herself to maximize purr decibels received. And this "solicitation" purr was not the agreeable thrumming of a content cat but an insistent chainsaw br-rr-oom demanding attention.

When the recordings were played to fifty volunteers, the solicitation purr was consistently rated as "more urgent and less pleasant."

The researchers then went back to the recordings and analyzed the acoustic properties of the purrs. The most consistent difference was the presence in the solicitation purr of a high-pitched component not present in the contentment purr. To test whether this element truly was responsible for the different perceptions of the observers, the scientists digitally manipulated the solicitation purrs by deleting that component. When a new batch of observers were played the original and digitally altered versions, they rated the manipulated ones as much less annoying.

At the end of the scientific paper reporting these results, the scientists suggested that the acoustic structure of the solicitation purr shared similarities with the cry of a human baby. Humans are known to be particularly sensitive and responsive to this sound; the scientists suggested that cats had evolved to tap into our preexisting sensitivity to create a purr that would get our attention.

When I read this, I thought it was nonsense. It's one thing to do a fancy-pants statistical analysis to detect similarities on a computer, but just because the properties of a purr and a crying baby share some digital similarities doesn't mean they really sound like each other.

Then I listened to the audio files the authors made available online with their paper. And what do you know? When I played the recordings, I could hear the similarity to a crying baby! I then found some other solicitation purrs on the internet, and if I aurally squinted really hard, I perhaps could hear a baby's cry in them as well.

All small-cat species can purr, so the ability to purr must have evolved long ago, well before domestic cats began associating with humans. Indeed, zookeepers have reported friendly felines of several species purring in their presence. But just as with meowing, domestic cats seem to have evolutionarily modified their purrs to better communicate with us.

This conclusion, however, assumes that an African wildcat—or even an ocelot or bobcat—wouldn't give an insistent, human-baby-like purr when waiting to be fed. If this assumption is wrong, then the similarity to a human baby's cry would have to be a coincidence, rather than an evolved means of manipulating us, because other small-cat species didn't evolve in the presence of humans.

As far as I'm aware, there are no published data to test this assumption. To dig further, I spoke to a number of small-cat keepers from zoos around the United States but was not able to get relevant information. A number of species are kept as pets, so maybe someone knows the answer, but that someone is not me.

My hunch—based on no evidence—is that even if tame ocelots, for example, purr when they hope to be fed, those rumblings are unlikely to sound like wailing human babies. Still, without the data, who knows? Seems like a good topic for a master's degree project!

PEOPLE HAVE LONG SUGGESTED that cats have us wrapped around their little toes, masterfully manipulating us to get what they want. The data on meowing and purring show this idea to have an evolutionary basis. Communication, however, is just one aspect of cat-human interactions. Given all the other amazing behaviors in the housecat's repertoire, surely they have evolved in other ways while adapting to living with us.

Three

Survival of the Friendliest

As I sit in my reclining chair typing these words, my trusty sidekick Nelson is sprawled across my torso, his brown foot occasionally contacting the laptop's touchpad and introducing new approaches to spelling and punctuation. Many of you no doubt have similar relationships with your feline companions. Such intimacy would seem like the height of domestic bliss. How can people say cats are only "barely domesticated"?

The friendliness of many (though certainly not all) housecats would seem like strong evidence of the fully domesticated state of housecats. What other feline species will cuddle in your lap, lick your hair, or follow you around the house? The underlying assumption has been that the housecat's ancestor was not friendly at all to humans (it's called the African "wild" cat for a reason) and that Chester's cuddliness is a recent evolutionary development, the result of domestication.

But you know what happens when you assume. What we really need is firsthand knowledge of the amiability of wild felines compared to our housemates. And who would know better than the zookeepers who work with these species every day, even entering the cage with smaller species?

To tap into this knowledge of cat affability, a behavioral scientist surveyed keepers at seventy-one zoos. Data were compiled for nearly four hundred small cats. Not surprisingly, the responses revealed that feline species vary substantially in temperament. Some species will sit or roll on their back next to their keepers, even rubbing against or licking them. Others won't have anything do with their caretakers.*

South American small spotted cats—the ocelot, Geoffroy's cat, and margay—won honors as the friendliest felines. Beautiful spots, however, are not a sure sign of camaraderie: the keepers reported the most unfriendly grimalkin to be the Asian leopard cat,† a distinction that will take on greater significance later in this book.

A Geoffroy's cat.

*Technically, the researchers recorded what they termed "affiliative" behaviors, which are behaviors "relating to the formation of social and emotional bonds with others or to the desire to create such bonds." The affiliative behaviors were: "sitting within 1 meter of the keeper; rolling within 1 meter of the keeper; head or flank rubbing on the keeper; and licking the keeper."

†Note: leopard cats are the size of housecats, but with the spotted coat of a leopard, hence the name. The two species are not closely related.

Even some closely related cats differ in disposition. Among the friendliest is the African wildcat, and among the unfriendliest is its close relative, the European wildcat (historically, African and European wildcats, as well as those from Asia, have been considered members of the same species; more on this later, but keep in mind that when I use the term "wildcat" I am not referring to just any wild-living feline, but to members of this specific species). These findings correspond with many reports from people who've kept these animals. African wildcats, if raised from kittens, are said to develop into affectionate companions, whereas despite the most tender attention, European wildcats grow up to be hellaciously mean.

Evidence of feline friendliness doesn't come just from zoos. People have reared young felines of different species in their homes with the goal of keeping them as pets. It turns out that many cat species, if raised appropriately, will make enjoyable companions. Even big cats can be kept as pets if raised carefully—mountain lions are said to make particularly good houseguests (not that I'm encouraging the practice; keeping mountain lions* and other wild cat species as pets is a bad idea for many reasons).

This habit of rearing wild felines has a long history. The ancient Egyptians, for example, tamed not only African wildcats (which they subsequently domesticated), but also cheetahs, lions, leopards, jungle cats (long-legged, short-tailed tawny cats with a long muzzle), and servals (beautiful, spotted, extremely long-legged felines from the plains of Africa—more on them later). In total, people have tamed fourteen species of cats over the last several millennia, mostly in Africa and Asia. Overall, an almost perfect correspondence exists between the

*There are many other names for the mountain lion, puma and cougar being the most common.

cats that zookeepers reported to be friendly and those that historically have been kept as tame animals.

There is, however, an important distinction between a tame animal and a domesticated one. The difference is an example of nature versus nurture. Tame animals are biologically no different from wild members of the same species, but they behave differently simply as a result of how they were raised. Bring up a female mountain lion in a house around people and she will be friendly. Release her into nature and her kittens will be as wild as any other mountain lion. Domesticated animals, in contrast, have evolved genetic differences that make them different from their wild ancestors.

So where does that leave our household friends, tame or domesticated? The answer is: a little of both. Housecats aren't that different from mountain lions. Kittens require human contact to develop into friendly cats. If they don't grow up being handled by humans, perhaps because their mother is feral, most become irreversibly wild. Four to eight weeks of age is a critical period; kittens handled regularly in this interval grow up to be well-adjusted housecats. By contrast, kittens first handled at eight weeks often end up being more reserved, and those not handled until ten weeks of age will rarely become friendly to people, no matter how kindly they are subsequently treated.*

On the other hand, housecats that interact with humans as kittens often grow up to be more friendly than other, similarly raised feline species. An ocelot is not going to sit on your lap as you tap away on your laptop, nor are you going to walk around the house with a limp serval in your arms regardless of how well they are socialized while

*Puppies, too, have a critical window of socialization, but it is later, when they are seven to fourteen weeks old. Kittens do better when they are handled by multiple people early on so that they learn to be friendly to all people, rather than to just a particular individual.

growing up. Housecats are not simply tame African wildcats. They have the capacity, if raised in the right environment, to be more friendly and affectionate than other felines. That congeniality enhancement is the result of evolutionary change during the domestication process.

This conditional amiability of housecats illustrates why nature versus nurture is a false dichotomy. An organism's behavior is a result of the interaction of nature and nurture. Neither the right genes nor the right environment alone is sufficient. Rather, it's the combination of the two that produces ultra-friendly cats.

Still, this behavioral difference between domestic cats and their wild cousins is relatively modest. Consider how greatly transformed dogs are by domestication: wolves, no matter how they're raised, are nothing like their slavishly obedient, owner-adoring descendants. Compared to the wolf-to-dog transition, the difference between African wildcats and domestics is much less dramatic. The terms "barely" or "semi-domesticated" are just matters of opinion, not scientific findings, so you'll have to form your own opinion about where on the domestication spectrum cats lie. To help you do so, let's consider other ways in which the behavior of *Felis catus* has changed from his ancestor.

ONE DAY SHORTLY after he joined our family, Nelson walked into the kitchen with one of my wife's cashmere gloves in his mouth. Why the kitten had taken a liking to it we have no idea, but he dropped it at my feet. If I picked it up and waved it in front of him, he swatted at it; after snatching it from me, he would roll on his back and attack it mercilessly with every sharp point on his body. When I tossed the glove across the room, he madly dashed after it, immediately bringing it right back to me. This occurred repeatedly for months. As the mitt became increasingly shredded, Nelson expanded his retrieval playlist to cat toys and other objects he found enchanting.

I was astonished. We already considered Nelson a dog in cat's clothing due to his friendliness and affection, a trademark of the European Burmese breed to which he belongs. But this was extraordinary: a cat that fetches! None of the previous seven cats with which I had lived had done this, nor had I ever heard of such a thing. I wasn't the only one. In 2019, NPR proclaimed in a headline "Cats Don't Fetch."

Nelson says "play time!"

Truly, I thought, Nelson is amazing, the most wonderful cat in the world. My head filled with fabulous ideas: a national tour, *The Tonight Show,* the Nelson YouTube channel. Fame and fortune to follow.

But then it occurred to me: perhaps I should check just to make sure that Nelson's abilities are truly unique.

Some quick googling set me straight. The internet is awash in videos of cats fetching toys. There's even a small amount of literature on the subject. One online survey of nearly three thousand cat owners revealed that twenty-two percent of cats brought toys to their human companions to initiate play. Another poll of more than four thousand Finnish cats revealed that fetching behavior was common, with Siamese cats taking top honors.

I have no doubt that when Nelson carries a toy to me and drops it at my feet, he wants to play. Many animals, wild and domesticated, engage in play, particularly when they are young, though some researchers suggest that domesticated species do so to a greater extent. Why animals play—perhaps to develop motor skills, learn how to interact socially, or practice hunting—is a topic of considerable academic discussion. Certainly, the common play behavior of kittens—stalking, pouncing, and wrestling matches—could serve these purposes and likely occurs in all feline species. By contrast, it's hard to imagine that fetching is a behavior that evolved in wild cat species.

Of course, the question is not whether wild cats in nature will fetch a toy if you throw one—of course they won't. They'll just run away, or eat you if they're big enough. Rather, the question is whether fetching behavior is latent in other species so that tame individuals will exhibit such behavior. The alternative possibility is that wild felines, tame or not, won't fetch and bring toys, which would indicate that fetching behavior evolved during domestication of the housecat.

Distinguishing between these alternatives requires examining the behavior of tame members of other cat species. In talking with zookeepers, I was not able to settle the question. Some cats can be trained to fetch, but that's not the same as a cat displaying such behavior spontaneously.

In fact, housecats do a lot of things that seem like they might be the result of evolution around humans. Consider making biscuits. If you live with a cat, you may have experienced the delight or aggravation (or both) of Mr. Patches standing on your belly, rhythmically pushing down with one forepaw, then the other, a vacant look in his eyes. Seemingly in a trance, cats may continue for many minutes before settling down for a nap.

Kneading is a behavior that kittens exhibit while nursing, presumably to promote the flow of milk from mama cat. Scientists and cat enthusiasts have long speculated why adult cats comfortable in the presence of people retain this behavior and direct it toward us. All agree that in some way it exhibits contentment, but the actual explanation for why such juvenilization* occurs is unclear.

Wild feline species knead as kittens, just like housecats. But then they stop. Or so said the zookeeper survey of four hundred–plus felines. But not everyone agrees. Some webpages state, without documentation, that in many or all wild species, some adults engage in kneading behavior. I wouldn't give these statements too much credence—you can find someone saying almost anything about cats somewhere on the internet—except that several knowledgeable people have told me that adult hand-raised African wildcats, ocelots, and other species will knead. Clearly, another dissertation project waiting to be conducted! Nonetheless, the data currently in hand suggest that adult kneading as a sign of contentment in the presence of humans may be a housecat adaptation to living with us. Why it evolved is unclear. Perhaps in some ways it builds stronger bonds or otherwise induces people to treat cats more nicely.

Other behaviors have received even less study. Research has demonstrated that when cats are confronted with a potentially scary, unfamiliar situation—like long green ribbons fluttering in front of a fan—they look to familiar humans for guidance about whether to be afraid. Cats also can recognize their names, read and respond to the emotional state of their humans, and follow people's gazes and finger pointing to find food (although I've tried that on my cats without success). These seem like behavioral traits that would have arisen as cats evolved to interact and live with humans.

Before we get carried away, let's think of some of the other zany

---

*"Neoteny" is the scientific term for the evolutionary change that leads to the retention of juvenile characteristics into the adult life phase.

behaviors we associate with housecats: they chase laser-pointer dots, sit in boxes, and get high on catnip. These also might be thought of as behaviors evolved by housecats. But my unscientific sampling of YouTube's offerings suggests that most species of feline—big or small—enjoy box sitting and are entranced by catnip. Response to laser beams is less consistent, but many wild cats go as crazy chasing the little red dot as any housecat I've seen.

Comparison between our domestic friends and their wild relatives is an area, sadly, where canine scholarship has outpaced work on felines. Scientists have demonstrated, for example, that dogs are better than wolves at following the gaze of humans to find something they want (like a tossed ball). Similarly, a dog staring into the eyes of a familiar person experiences a surge of oxytocin (the "love hormone"); not so even for hand-raised wolves.

Other than Nicastro's work on African wildcat meowing, comparable studies have not been conducted on other feline species. Clearly, more research on domestic cats and detailed behavioral comparisons with African wildcats and other feline species are needed to understand which traits are unique to housecats and likely evolved as they became our domestic companions.

So far, the behaviors we've examined involve how cats interact with humans. However, there is one behavior that domestic cats exhibit that is nearly unique among felines and thus a trait involved in domestication, yet is directed to other cats as much as it is to humans. Nelson once again will introduce the story, and I can guarantee it's an uplifting tale!

Despite our best intentions to raise Nelson as an indoor cat, he is desperate to explore the great outdoors. Occasionally, we give in and let him out into the backyard with a cat tracker attached to his collar so that we can find him if he goes over or under the fence.

Sometimes he does, and then I have to track him down and bring him back. When I locate him, he at first doesn't seem to recognize me and appears apprehensive (cats' distance vision is not as good as ours, though, of course, in the typical cat way, maybe he's just pretending he doesn't know me). But as I get closer and call to him in my best "Nelson, buddy" voice, he eventually starts walking, or sometimes running, to me. As he gets closer, his tail springs straight up into the air, a rear-end exclamation point; when he gets to me, he rubs his cheeks and flank against my leg, purring all the while. Sometimes he behaves similarly inside when he's in a loving mood, approaching with tail held high, then licking my hand or foot in exchange for caresses, sometimes even rolling on his back for belly rubs. Most people who live with friendly cats are familiar with this behavior sequence.

Domestic cats use this same flagpole signal when they interact with each other: the upright tail means "I come in peace" or maybe "Glad to see you!" An approaching cat raises his tail as a signal that he wants to engage in other friendly behaviors like head and body rubbing, nose touching, and sniffing; the other cat returns the vertical salute to indicate he is receptive to such an interaction.

A friendly cat with tail up.

Behavioral scientists tested the communicative value of an upraised tail in a laboratory study. It's long been known that cats initially react to anatomically correct silhouettes as if they are real felines (though they quickly figure out the ruse). Taking advantage of this knowledge, the researchers pasted a cat silhouette on a wall and then introduced a real cat into the room. When the cutout had a tail held high, the pet cats frequently raised their own tails and approached the silhouette quickly. In contrast, when the shadow's tail was down, the real cats raised their own much less frequently and took more than twice as long to approach. The cats also wiggled their tails five times as much when viewing the tail-down silhouette, indicating uncertainty and nervousness.

Clearly, a tail held high is a friendly cat message. The fact that cats use their tails to signal amiable intentions to us as well is a great tribute, indicating we've attained honorary cat status.

Only one other feline species uses its tail in a similar way. Surprisingly, it's not another small feline species, but, rather, the King of the Jungle. When greeting each other, members of a lion pride will raise their tails—though in a more curved semicircle than straight up—while rubbing their heads or smelling their rear ends. Domestic cats and lions seem like an unlikely pair of felines to share this unusual behavior. In fact, however, there is a ready explanation, one that highlights the most significant evolutionary leap domestic cats have made from their wildcat ancestors.

Four

Strength in Numbers

A common characterization of pets is that dogs are loving, gregarious, social animals, whereas cats are aloof loners. This difference makes sense given that dogs are descended from a species that lives in packs, the wolf, whereas cats come from a line of species commonly thought to live on their own. However, as we'll see in this chapter, the social lives of domestic cats—and to some extent their larger relatives—are much more complex than commonly realized.

Lions, of course, have always been known as the exception to feline asociality, famous for living in prides composed of as many as twenty-one related females (though five is a more typical number). Pride members live in association with one to several males (rarely as many as seven) that are unrelated to the females. The strong social bonds that tie a pride together are evident in the endearing displays of affection by pride members, who rub up against, groom, and lie on top of each other.

The social nature of lion prides is also evident in how they hunt cooperatively, working to bring down prey too large for any one lion to kill, sometimes even a giraffe or a midsize elephant. These are true coordinated operations, rather than just several lions independently prowling in the same area. The lions use complex strategies and coordination, such as several lionesses driving prey in the direction of hidden compatriots lying in ambush.

Social interactions extend beyond hunting, permeating all aspects of pride life. Females that have cubs at the same time will jointly raise them, nursing each other's cubs and leaving one lioness behind to mind the crèche when the rest go out on a hunt. Pride members also work together to defend their territories against other prides.

By contrast, tigers and leopards spend most of their time living by themselves or, in the case of females, with their young. When individuals encounter each other, interactions can range from highly aggressive, to nodding as they pass by, to—at least in tigers—feeding more-or-less peacefully from the same carcass. Unlike lions, neither tigers nor leopards stay together in groups, nor do they hunt, defend food or space, or raise their young cooperatively.

For the most part, all other cat species follow the tiger-leopard model. That statement, however, requires two caveats. First, the natural history—what species do in the wild—of most small feline species is poorly known. We don't have a lot of details about the lives of many of the more obscure cats. What we do know suggests that they are all similar, solitary species, but who knows what surprises may turn up once someone studies them in detail?

The second caveat pertains to the most unusual of cats, the cheetah. Distinctive in their extremely long legs, which provide the ability to sprint at seventy miles per hour, and in their doglike non-retractable claws, these spotted speedsters are unique in their social organization as well. Like most felines, adult females spend their lives alone or with their cubs. But the males are a different story. Coalitions, often composed of brothers, band together to control a territory and mate with the females that live within it. Like lion prides, cheetah coalition members sometimes hunt cooperatively, but they only associate with females during courtship.

Wild cats thus exhibit true sociality (lions), semi-sociality (cheetahs), and asociality (all the rest). Where do domestic cats fit in?

WE'LL START WITH the easy part of the answer. No parallel to cheetahs exists—no one has ever reported brother toms banding together to attempt to control mating access to female cats. In fact, curiously, the opposite happens. Although male domestic cats can be very aggressive toward each other, they are surprisingly tranquil when courting a female in heat.* Instead of one male fighting off other suitors, toms will often remain calmly in the vicinity while the female mates with multiple males.†

When domestic cats live in the wild, away from humans, they usually live solitary lives, adults rarely encountering each other. Such cats roam over large areas (a topic we'll discuss at greater length in chapter sixteen). Sometimes, individuals will have exclusive prowling grounds, but more commonly, multiple cats will travel over the same area.

---

*Technically, "in estrus."

†Lions are famous for the frequency with which they mate when the female is receptive—as many as fifty times a day for several days according to some reports—but female housecats are almost their equals, copulating fifteen to twenty times per day for four to five days.

These cats, however, rarely meet each other. In part, that's because they cover a lot of ground, so the chance of bumping into another cat by accident is relatively small. But just to be sure, they leave calling cards advising others to keep away. Cats have an excellent sense of smell, and their primary means of signposting is to leave stinky messages, emanating from strategically placed feces or urine, to announce their presence in an area. As a result, it's inaccurate to say these cats are asocial. Thanks to chemical communication, they interact a lot, just not while in each other's presence.

When they do run into each other, however, their interactions often are not amicable (with the exception of a courting male approaching a female in heat, though those meetings can also be feisty). The scientific literature contains surprisingly few reports of what happens when two feral cats come into contact. In one study in the wild in the Galápagos Islands, researchers observed fourteen feral cats for more than two hundred hours. During that time, two cats were observed coming together forty times. When the cats were both males, they "engaged in mutual olfactory inspection of nose, shoulders, and anal region followed by low, throaty vocalizations. The climax of the encounter was characterized by a high-pitched vocalization accompanied by a quick blow. The subordinate individual lay on its side while the dominant individual stood across its head in a stereotyped, stiff-legged, arched-back posture continuing the low vocalizations. After a short time, usually less than a minute, the dominant individual departed; the first few steps were stiff-legged. After a brief interval the subordinate individual arose and departed, usually in the opposite direction. No amicable encounters or assemblages of males were seen."

The solitary lifestyle of feral cats parallels that of almost all feline species. But things are different when cats live around us. In many places, large numbers of outdoor cats live in the vicinity of humans, subsisting primarily on discarded food or handouts. These clowders* were long considered nothing more than aggregations of a lot of cats with no particular social tendencies. However, once researchers started conducting detailed studies on cats on farms and in urban localities where they are fed, they realized that domestic cat colonies are much more than a bunch of cats that happen to live in the same place.

Rather, colonies are often divided into subgroups, each composed of related females. Members of a subgroup are amicable with each other, while often behaving aggressively to other females in the vicinity. Kittens are reared communally, a queen nursing any hungry kitten from the subgroup.† Females will even assist others in giving birth, essentially serving as a midwife. One queen, for example, was observed severing the umbilical cord and cleaning another female's newborn kittens. Large colonies of cats have many subgroups, the largest, most powerful subgroups located near the central source of food (if there is one) and smaller subgroups on the periphery.

The social structure of high-density domestic cat colonies thus shares many similarities with lion prides. In both, groups are composed of related females that stay with their family when they mature; males leave the group and move elsewhere to try to reproduce; and mothers jointly raise cubs born at the same time, even nursing each other's offspring.

The similar group living of domestic cats and lions explains why the two species, and no other felines, exhibit the tail-up display. It's not surprising that cats living in close proximity have developed a way of

---

*The term for a group of cats. I don't know why we don't call a group of domestic cats a pride, nor where the term "clowder" comes from (try looking into it and you'll find contradictory explanations).

†"Queen" refers to a female cat that has or has had kittens, or according to other definitions, a female that has not been spayed.

visually signaling friendly intentions. And what better piece of anatomy to use than the tail—easy to move, visible at a distance, and not tied up in other activities? Think of the alternatives. Legs would work, but they're often busy supporting the cat or moving her around. Ears do the job, but are harder to see at a distance, whiskers even more so. The tail is the perfect body part for feline visual communication. Its independent recruitment for social signaling by both domestic cats and lions is an example of adaptive convergent evolution, similar features evolving independently in species that experience similar situations.

Before getting too carried away, I need to point out that the social organization of the two species is not identical; there are some differences. For example, female housecats give birth in the presence of other group members, whereas lionesses head off into the bush and don't return until their cubs are six weeks old. Also, housecats don't generally hunt cooperatively (thank goodness—just picture a pack of housecats taking down groundhogs and raccoons!).

A major difference is the social organization of males. With lions, males band together and take control of a female pride, or occasionally more than one pride at the same time. Coalition members are often relatives who left their natal pride together, but not always.

Unlike lions, male cats do not form coalitions, nor do they usually restrict their attention to a single subgroup of females. Rather, most males cruise around, attempting to mate with as many females as possible.

These differences notwithstanding, the social organization of lion and cat groups is strikingly similar and greatly different—as far as we know—from all other feline species (with an asterisk for the cheetah). But how can we explain this convergent behavioral evolution in cats and lions and the lack of comparable sociality in other felines? Abundance is the key.

. . .

RESEARCHERS HAVE STUDIED unowned cat populations around the world, from the streets of Brooklyn to frigid islands near Antarctica. They've found that the density of cat populations ranges from two and a half cats per square mile in some places to more than six thousand per square mile in others. In case you're wondering, six thousand cats per square mile is about nine cats per acre or one per basketball court–size piece of land.

The cause of this more than two-thousand-fold difference is simple: food availability determines cat numbers. When cats are living on their own, not being provisioned by humans, they have to find food for themselves. In most places, prey aren't very abundant, and as a result, cats are scarce; it takes a lot of land to provide enough food for a single feral cat. The density of cats in such areas is low, about two to fifteen cats per square mile.

In contrast, on islands lacking native predators, prey populations can reach enormous densities. Seabirds, in particular, choose predator-free islands to establish their colonies, which can often grow to huge size. When cats are introduced, the seabirds are easy pickings. As a result, cats prosper, and their populations can grow to large numbers, at least until they wipe out the seabird population.

Many unowned cats, however, don't have to live off the land, at least not entirely. Cats living on farms prey on the abundant rodent populations, but they also receive handouts from the farmers (gotta keep the mousers happy!). As a result of this ample supply of food, farm cat populations are much larger than most unprovisioned cat populations.

Extremely dense cat populations, however, only occur in urban areas where immense amounts of food are available either because people intentionally set it out for the cats or because our waste is plentiful and available.

One such place is the old neighborhood of Nachlaot in central Jerusalem. Wikipedia tells us that Nachlaot occurs "outside the walls of the Old City" and is "known for its narrow, winding lanes, old-style housing, hidden courtyards and many small synagogues." The wikipage goes on to mention that "at one time Nachlaot had a higher concentration of synagogues than anywhere else in the world, around 300 within a radius of just a few blocks." But the internet's encyclopedia fails to cite an equally important claim to fame: the neighborhood boasts the highest concentration of cats ever recorded anywhere in the world.

We know this thanks to an Israeli graduate student named Vered Mirmovitch. Having grown up on an Israeli kibbutz where children were raised communally rather than in the homes of their parents, Mirmovitch was curious about how social systems evolve. During her college years in the early 1980s, she could only afford a subground apartment (sometimes called an "English basement") in Nachlaot. Looking out her window at street level, she had a perfect view of a dumpster across the street (except when a car parked in the way). Day after day, she watched cats interact as they fed on trash scraps.

Once she came to recognize individuals, she realized that the cats knew each other and lived in groups, behaving amicably to fellow group members, but with hostility to outsiders. Counter to the common wisdom that cats were loners, this observation sparked her curiosity, so she decided to make the cats the subject of her master's degree research.

Wandering the neighborhood, she quickly learned that the cats subsisted primarily on garbage bins, either the type many people wheel out to the curb on trash collection day or the much larger dumpsters.

In the Nachlaot neighborhood, however, residences didn't have their own bins. Rather, bins and dumpsters were located at nine places in the six-acre neighborhood. Every day, residents would take their trash out, usually after dinner, and walk to the nearest receptacle. And the cats would be waiting, maybe three or five at a trash bin, as many as a dozen at a dumpster.

The easiest pickings came from the people who put their bags next to the bins or even tossed the tastier scraps onto the ground. Even trash deposited in the receptacle was readily pillaged if the lid of the bin or dumpster was left open or if there was so much garbage that the lid couldn't shut.

Closed repositories posed more of a challenge. One cat, however, learned how to push open the lid to a garbage bin and wiggle his way inside. Then—because who wants to eat in an enclosed garbage bin?—he would jump up and hit the lid so it would swing open, providing access to all. The other cats would wait around for him to work his magic, but none ever learned his trick.

This plenitude supported a large population of cats, so many that it was hard to keep track of who was who. If she was going to study interactions among the cats, Mirmovitch needed to be able to identify all the players. To do so, she borrowed her sister's camera, snapping photos of every cat she saw and carrying around an album of mug shots. Quickly she learned to recognize them all by sight.

Eventually, she stopped finding new cats, indicating that she had them all in her registry, all sixty-three of them. And that's how we know that the Nachlaot cat population was exceptional. The sixty-three cats translate into a density of sixty-three hundred per square mile. To this day, this is the highest density of housecats ever recorded.

Mirmovitch's behavioral observations revealed a social organization that we now know to be typical for dense cat populations. Once she could recognize the cats, she could record who was interacting with whom. And what she discovered was that cats lived in groups and fed in circumscribed areas; each group primarily used one trash receptacle, but had a second, backup site that would be visited occasionally. Group members were lovey-dovey with each other, grooming, rubbing, and sleeping together, in sharp contrast to the hostility they displayed when a non-group-member showed up at their trash can.

A few years after Mirmovitch's work, Japanese researchers began to study cats in a very different setting, yet one that shared a common feature: a huge amount of human-provided food.

Ainoshima is a small island off the southwest coast of Japan. Most of the island is covered with grasslands, fields, and forest, but in the southwest corner is a village, about three times as large as Nachlaot and occupied by fisherpeople and cats. Rumor has it that cats were introduced to the island to keep rats from gnawing holes in the fishing nets. Whether they kept up their end of the bargain is unclear because the cats had another source of food: the piles of fish scraps that the fishers left daily at six points along the seashore.

The cats feasted on this bounty and proliferated. Recognizing the cats by their distinguishing marks, researchers identified two hundred cats, equal to sixty-one hundred cats per square mile, only slightly less than the density in Nachlaot. The social organization of the cats into groups that each fed almost exclusively at one of the dumps was also very similar to the Israeli cats.*

Nachlaot and Ainoshima are extremely different, but they're alike in one important way: in both places, cats are extraordinarily abun-

---

*To promote tourism, Ainoshima has been rebranded as Cat Heaven Island, one of the dozen or so Japanese cat islands renowned for their large populations of felines, sometimes exceeding the number of humans on an island.

dant and have very similar living arrangements. This highlights that it is the availability of food, and not some other factor, that drives the formation of social groups in domestic cats.

STILL, WE'RE LEFT with a question. Abundant food understandably leads to a concentration of cats, but that doesn't explain why the cats live in social groups. What's the advantage of forming these cliques, as opposed to continuing to live life as an unsocial loner, albeit one who bumps into neighbors quite frequently?

Keep in mind there's a cost to being hostile—you could get hurt—plus there's no reason to defend food when there's enough for all. Consequently, it's easy to imagine that living in places with abundant food selects for less aggressiveness. Still, that's not the same as being positively friendly and behaving cooperatively with nearby cats.

To answer this question, let's start with lions. For decades, researchers have asked the same question about them: why do lions live in social groups, instead of singly, like female cheetahs which occur in many of the same places on the African savannas?

Many ideas have been proposed. Initially, the preferred hypothesis was that group living enhanced hunting success: bigger groups were postulated to be better at catching medium-size prey, like zebras and wildebeest, and more able to tackle larger prey too big for single lions to catch on their own.

Both of these premises are correct: groups have higher success rates than individuals hunting alone and can even occasionally bring down a midsize elephant. But there's also a drawback to living in groups: there may be more food, but there are also more mouths to feed. Indeed, when scientists did the math, they found little advantage in group living—the amount of meat per animal is no greater, and is often less, in larger groups.

So, if it's not for the eats, what's the advantage to group living? In a word: defense. In open settings like the Serengeti Plain, there are no secrets. When a successful hunt occurs, other animals will see it, and the presence of circling vultures will transmit the news for miles. A single lion can be pushed off a kill by a group of hyenas. The greater the number of lions in a pride, however, the less often large packs of hyenas will be able to drive them away.

Living in groups thus allows lions to retain their kills after they make them. And more generally, large prides get the best territories and can keep other prides from encroaching on them.

There's one more reason that lionesses live in groups. It revolves around the darker, seamier side of lion life, so avert your eyes (or skip to the end of the next paragraph) if you're squeamish.

Here we go. When a new group of males takes over a pride, they kill all the young cubs. That sounds horrible, but the behavior—which occurs in other species as well, including some monkeys—has its evolutionary logic. Why should a male lion expend energy raising another male's offspring? With their young gone, females will more quickly come back into breeding condition and more quickly bear the offspring of the new males. This is not a trivial consideration. On average, males control a pride for only two to three years, so there's no time to waste in sowing their oats.

It's unpleasant, but that's nature. Any trait that enhances an individual's ability to pass on its genes to the next generation will be favored by natural selection, whether that trait is longer legs, bigger brains, or killing the offspring of rival males so that females will bear your progeny.

The relevance of male infanticide to group living is that when invading males encounter a single female with cubs, all the cubs are almost always killed, but when males meet a group of two or more lionesses, usually at least some of the cubs survive.

So group living in lions comes down to defense: defense of territory, defense from predators, defense from invading males. Does the same explanation hold for domestic cat group living?

To a large extent, it does. Communal care boosts survival for kittens just as it does for lion cubs. Multiple queens working together can better raise youngsters, both because one queen can leave to go foraging while another tends to the home front and because multiple queens can better defend against predators, dogs probably being the biggest threat.

In addition, infanticide by tomcats, though not nearly as common as with lions, does occur.* And just as with lions, multiple female group members are probably more effective in detecting and driving off marauding males.

The dominance of larger groups of domestic cats is another parallel to lions. In locations where a large amount of food is deposited in a small area—as often happens with garbage piles or caretaker feeding sites—the largest groups of females live closest to the food. Females from other groups are not excluded, but they certainly don't get a warm welcome as they intrude on the larger group's space to get to the food.

One difference with lion prides, though, is that group-living cats are not known to work together to defend their food from other predator species.

---

*Though remember that groups of toms don't have exclusive mating rights with a group of females like coalitions of lion males do, which perhaps explains why infanticide is much less common in domestic cats.

Overall, domestic cats and lions are cut from the same cloth when it comes to their social lives in the wild. High food abundance leads to high population density, which in turn leads to the formation of groups of related lionesses or queens. These females work together to claim the best areas in which to live and defend their young and their resources. The evolution of this capacity for social living in domestic cats is the single most significant evolutionary change in their descent from African wildcats.

IF SOCIALITY IS SO ADVANTAGEOUS to domestic cats and lions, why don't other felines also live in groups? Presumably, the answer is scarcity of food: most habitats are not as bountiful as the plains of the Serengeti or the garbage cans of Nachlaot. And without much food, there aren't enough cats around to make group living feasible.

This explanation may well be correct, but it's actually very difficult to quantify how much potential food is available for a particular species in a particular place. Nonetheless, researchers have been able to show for a few species—Eurasian lynx, ocelots, and lions—that areas with more prey have larger feline populations.

For most species, we don't have such information, but we may not need it. If cat populations are very sparse—regardless of the reason—then we wouldn't expect the formation of groups. To evaluate that proposition, all we need are estimates of population size, information that is more readily obtained than data on prey availability. Our prediction is clear: most feline species should have low population densities compared to those of domestic cats and lions.

Estimating the population size of ocelots in a tropical rainforest is a lot more difficult than doing the same in a Jerusalem neighborhood. In a small, open area, it's possible to observe and learn to identify every animal that lives there, as Mirmovitch did. But for species that are scarcer, scientists need to work over a much larger area, which in turn makes it much harder to find every individual.

To get around this problem, scientists use statistical methods to estimate what the population of a species is in a given area. I won't go into the gory mathematical details—suffice to say there are algorithms that use the information on how many of the individuals have been seen five times, how many four times, how many three times, and so on to infer how many more there are that haven't been seen at all.

There's a second problem. These methods rely on actually seeing the cats. But wild felines are notoriously secretive. In all my time in the rainforests of Central and South America, I've only seen one cat, an ocelot. No jaguars, no jaguarundis, no margays. Thanks to recent technological innovation, however, scientists can get around this problem.

About the size of a thick paperback novel, a camera trap* takes photos and videos. Triggered by either movement or body heat, the camera snaps several stills or a short video, stores the images on a digital memory card, turns itself off for a few seconds, then waits to be triggered again. Researchers place camera traps in the wild, usually strapped to a tree at a height appropriate for the target species (waist high for a deer, for example, or near the ground for a cat). Once the trap is placed, the researcher need only return periodically to download the images and change the batteries.

Scientists often put out a large number of camera traps across a study area for weeks to months. If they can identify individual cats in the photos due to their markings, they use the algorithms I just mentioned. If not, they have other statistical techniques in their arsenal.

*Sometimes called a "trail camera."

One way or another, the images provide the ability to estimate the size of the populations.

Of course, even when you set up a camera to detect cats, there's a lot of by-catch. I've learned this with camera traps in my own backyard. Opossums, groundhogs, bluejays, and deer mice show up regularly. Even the occasional deer or coyote makes an appearance.

These images can be very entertaining: squirrels caught in midair, their front legs extended Superman-style; a mama opossum with eight babies hanging on for dear life; two brazen raccoons caught in flagrante delicto making the next generation of little bandits.

And sometimes the images tell you something about the biology of the species under study. Occasionally, for example, cats are seen carrying prey in their mouth, which provides information on what they're eating. In one study, images showed a deer licking or nuzzling surprisingly tolerant cats for reasons unknown.

The advent of camera trapping has revolutionized the ability to monitor populations of secretive felines living in cluttered or remote

A serval.

habitats. As a result, scientists now have obtained population estimates for twenty-five small-to-medium-size cat species. The most abundant wild feline population ever recorded was in a thirty-square-mile fenced area surrounding a petrochemical plant in South Africa where two and a half servals occurred per square mile. Several ocelot populations were not far behind, but most species had densities well under one cat per square mile.

The petro plant concentration of servals is less than one two-thousandths of the density of cats in Nachlaot; Mirmovitch recorded more cats on a small garbage bin than the number of servals that occur in a square mile of South African grassland. No wonder, then, that domestic cats have a very different social organization than similar-size wild cat species.

Lions sometimes occur at a density greater than one per square mile, about the same as some of the densest small cat populations and much less than the density of unowned domestic cats. At face value, this might seem to contradict the high-density explanation for social living in domestic cats and lions.

However, lions and other big cats are enormous in comparison to small cats and need more of everything—food, water, shelter, space—and thus occur at lower densities. Moreover, big cats eat much larger prey than small cats. Because mice are much more abundant than, say, zebras, reliance on large prey by big cats leads to fewer cats per square mile.

For this reason, comparing lions to servals is an apples-to-oranges comparison. To understand lion sociality, we need to compare them to other large cats. And when we do so, we see that lions occur at much higher density than most of the others, no doubt a result of the cornucopia of prey on the African plains.

But there is one exception. Leopards also sometimes occur at high densities. Why, then, don't leopards live in prides? The answer here, too, probably comes down to defense. On the open African savanna, a pride of leopards couldn't stand up to a pride of much larger lions. So instead they live singly and drag their prey high into a tree, safe from earthbound lions and hyenas (cheetahs also suffer from losing kills to larger predators, but because they are not adept at climbing trees, their only recourse is to bolt down as much food as they can as quickly as possible).

ALTHOUGH OTHER FELINES don't have the advantage of living in our homes or off our handouts or garbage, many species do live in a variety of different circumstances. If food availability drives variation in domestic cat social arrangement, we might expect to see similar flexibility in other species.

Indeed, there are some hints that this may occur. Lion pride size varies in relation to food availability, being smaller in areas like the Kalahari Desert, where food can be scarce. There are even claims—not well substantiated—that the Barbary lion of northern Africa, now extinct, did not live in prides at all.

Similar tantalizing observations were made for the African wildcat. Generally described as living a solitary life, these secretive cats have been the subject of surprisingly little research. The only study to date involved radio-tracking six cats around a small settlement in the middle of Saudi Arabia that contained a wildlife research station, a dairy farm, and a camel milking station. The area had plenty of dumps and refuse sites where everything from picnic leftovers to goat carcasses was available in abundance. The wildcats visited regularly, but did not socialize with each other, much less form large colonies.* This observation suggests that African wildcats, in contrast to domestic cats, do not take advantage of abundant food resources to form social groups.

Nonetheless, there is the possibility that not all African wildcats follow the lead of the Saudi Arabian Six. A century ago, a British naturalist twice observed African wildcats living in close proximity to each other—potentially in a colony—in holes made by fennec foxes or other animals. In one case, the wildcat aggregation occurred in an area swarming with gerbils, suggesting that the abundance of prey was responsible for the atypical living arrangements. We shouldn't get too carried away by these century-old, detail-deficient reports. Nonetheless, they do raise the possibility that African wildcat sociality varies with food availability, just as in domestic cats and lions.

Because the African wildcat is the ancestor of the domestic cat, it is a shame that we don't have a better understanding of its natural history. The general assumption is that the ability to live in groups is a feature that domestic cats evolved in the last few thousand years as they began to live around us (more on this scenario in chapter seven). But if the African wildcat also has those tendencies, rather than social living being an evolutionary adaptation by domestic cats for living among humans, the causality may be the other way around: the propensity to live in groups when food is abundant may have been an important predisposition that paved the way for cat domestication. In the absence of better data, for the rest of this book, I'm going to stick with traditional wisdom that sociality evolved in domestic cats; none-

*One wildcat male did hang out around the Royal Pigeon House along with twenty feral domestic cats. Several times he was found sleeping snuggled up with several of them, and once he was even observed mating with a domestic female.

theless, we need to be open to the possibility that new information may change this understanding.

LEAVING ASIDE LIONS, most other feline species are thought to be solitary, only coming together to mate, and otherwise avoiding each other and not getting along well if they do meet. This may be true in many cases—ocelot males are known to kill other males, for example—but recent research has added a wrinkle. Some feline species that live most of their lives alone aren't necessarily unfriendly around food.

Mountain lions fit the traditional bill; other than mothers and their kittens, these cats are loners. Or so we thought. By using GPS radio collars to track the movements of mountain lions around Grand Teton National Park in Wyoming, researchers discovered that the cats commonly fed together at elk kills. All thirteen mountain lions in the study shared a meal with another cat at least once. Motion-triggered camera traps placed near carcasses recorded that the cats mostly interacted amicably, the occasional hiss or swat notwithstanding.

The size of the prey may explain this unexpected social tolerance. In this area, mountain lions mostly kill elk, which are much too large for one cat to eat, even over a number of days. The researchers speculate that the lions were behaving in a reciprocally altruistic way—you let me have some of your venison today, I'll let you have some of mine next week. Less detailed reports suggest that other large, normally solitary big cats—jaguars and tigers, specifically—also share large or abundant food.

This shared dining probably only applies to the sharing of very large prey by the large cats; it seems unlikely that rusty spotted cats share grasshoppers or that servals share mice. Still, who knows what surprises await when we learn more about the biology of small-cat species? Perhaps we'll discover similar unrecognized social tendencies that set the stage for the evolution of group living in domestic cats.

WHETHER THE AFRICAN WILDCAT had latent social tendencies or not, the rise of agriculture was a watershed moment that set cats on the path to our living rooms. The origin of permanent human habitations, combined with ample foods stored in central locations, led to a vast concentration of rodents and other potential prey. Cat populations boomed.

High densities around rich food sources guaranteed that cats would come into proximity regularly. Like mountain lions sharing an elk, the African wildcats had to be able to coexist without slaughtering each other. Fighting risks injury and is unnecessary if there's enough food for all. Consequently, as living peacefully at high density became the norm for these felines, they evolved to be better able to get along. It was probably for this simple reason that sociality evolved in domestic cat ancestors living around early agrarian communities. We'll discuss more about when and where domestication occurred in chapter six.

For dogs, which made the same transition a few millennia earlier, domesticity translated into even more sociality. Bring a new dog into a dog-full household and quickly they're all best buds. For cats, it's a different story. Living in our homes has not led to more advanced sociality.

Sometimes, cats in the same household get along well; other times, though, the atmosphere is less friendly and amicable than seen in outdoor cat colonies. I know this from personal experience.

My first cat, Tammy, was joined two years later by another shelter adoptee, Mauritia.* The two cats hit it off like gangbusters and for the next fifteen years routinely slept side by side, groomed each other frequently, and seemed to be best friends.

Tammy and Mauritia sleeping together.

Sadly, our current cats are a different story. When Nelson arrived a few years ago, the house was already the domain of Jane and Winston, siblings whose feral mother had been run over by a car when they were only two weeks old.† Fortunately, someone in the neighborhood knew the queen had just had kittens, located them, and turned the orphans over to a friend to raise. Four months later, they came to live with us.

W and J had the house to themselves for six years, and they were not happy at the arrival of a rambunctious youngster. All Nelson wanted to do was play, but they would have none of it. And the more they spurned him, the more aggressive he became in his quest for

*Pronounced "Mar-ish-a" and named for the Indian Ocean island of Mauritius that had attained near mythical status in my father's mind.

†Remember when I mentioned that females will mate with several males? Winston and Jane look nothing like each other. Winston is a large, seventeen-pound tom (all, or at least mostly, muscle!), gray blotches on white, with a big, solid head. Jane, on the other hand, is a petite eleven-pounder, slate gray in color. I always thought this odd, given that they are siblings, until I learned that it is common for feral female cats to mate with multiple males when they are in heat, leading to litters of half siblings; one study in a French city found that three quarters of litters had multiple paternity (in a rural area with lower cat densities, the proportion was substantially lower).

interaction. Now that Nelson is full grown, the relationship has become a permanent chill in which he bullies the smaller Jane and harasses Winston, who refuses to stand up for himself despite his much greater size.

The situation with Nelson is not unusual. In one survey, forty-five percent of respondents who live in multi-cat households reported fighting among their cats at least once a month.

One key to understanding what goes on in your home is to remember that social groups in cat colonies are usually based on family ties, often composed of several generations of female descendants of a single queen. In contrast, except when someone adopts littermates, most cats in a household are not related. Putting unrelated cats together can thus be tricky. As one cat behavior expert explained, the cardinal rule of cat society is: "proceed with caution when meeting any cat that has not been part of your (cat) family for as long as you can remember."

I would expect that cat fighting is more prevalent in households where the cats are not related, but I am unaware of any data testing this hypothesis. One study did find that when pairs of cats were left in a cattery while their owners were out of town, siblings lay next to each other, ate at the same time, and groomed each other much more often than unrelated cats that had lived together for years.

There's a second difference between unowned cats living in groups outside and pet cats living in a house with other cats. In outdoor colonies, one or several cats can drive another cat off, making life so unpleasant that she goes to live elsewhere. By contrast, cohabiting indoor pets have to carve up whatever space is available in the house in which they live. Homeowners can make the situation worse by not providing enough food, water, litter boxes, or napping locations; as a result cats are forced to trespass into the preferred areas of others, enhancing the likelihood of conflict between cats that already don't get along.

Although cat colony members are almost universally hostile to non-colony members, sometimes such interlopers, through persistence and determination, can become accepted into a colony. Getting unrelated cats to coexist peacefully is thus not an impossible dream. Bookstore shelves are lined with books by cat experts explaining how to introduce a new cat into the household, and many of these efforts, as with Tammy and Mauritia, are successful.

This success might suggest that, given enough time, cats will travel the route of the dog, becoming ever more socialized as the domestication process proceeds. At the moment, however, I must conclude that the label "semi-domesticated" is appropriate for cats, especially in comparison to the much greater transformation of the dog and other domesticated species. Group living is the major difference between non-pedigreed domestic cats and African wildcats. The other changes—minor alterations in anatomy and increased friendliness—are matters of degree. Non-pedigreed domestic cats just aren't all that different from their ancestors, a conclusion driven home by the ease with which feral cats revert to wild living.

Is this as far as it goes, or are cats headed for a more domesticated future? Before we can ask that question, let's consider the path they've already walked to their current state of domesticity.

Five

Cats Past and Present

Traditionally, the fossil record has been key to deducing how species changed and diversified through time. Paleontologists have taken this approach to study cat evolution, but they've been hampered by a surprising paucity of fossil felines: only sixty species are known over the thirty-million-year history of the cat family, barely more than the number of living species today. To make matters even more difficult, these fossils aren't evenly distributed: for some types of cats and some periods of time, we have a lot of fossils, for others, not so many.

The first cat, *Proailurus lemanensis,* was about the size of a bobcat. Short legs notwithstanding, it was clearly a cat. This, in fact, has been a hallmark of feline evolution—a cat's a cat. This may not seem extraordinary, but frequently it's not the case: for many types of animals, some extinct species were often quite different from their modern relatives. Consider, for example, that the giant ground sloth was nothing like the diminutive tree hangers we know today, lizards that looked like sea dragons grew to fifty feet in length and cruised the world's seas during the Age of Dinosaurs, and some ancient crocodiles lived on land and had hooves. In contrast, cats seem to have found a winning recipe and have stuck with it.

For the first ten million years, the feline fossil record is pretty thin: two more species of *Proailurus* evolved, but nothing more (of which we're aware). Then, about twenty million years ago, cat evolution kicked into gear. At that time the feline clan split into two branches. One of these gave rise to the many species of saber-toothed cats, which occurred throughout much of the world. We'll talk more about them later, but suffice to say, you would take note if one of these dental overachievers crossed your path (as no doubt happened to the earliest Americans, who arrived in North America before these cats went extinct). Still, enormous fangs and bulked-up forequarters notwithstanding, you'd have no trouble recognizing a saber-toothed cat as a feline.

The other branch of the evolutionary tree is referred to as the "conical-toothed" cats. This group, which contains all of today's feline species, is not well represented in the fossil record. Its species are unremarkably catlike and wouldn't surprise you if you saw one in a zoo.

Why we have discovered so many fossil saber-toothed species and so few of other types of cats is a good question. One possibility, of course, is that this disparity is a true record of what happened. Perhaps saber-cats were much more evolutionarily diverse than other types of felines, and perhaps the modern cat clan, so species rich today, only began to diversify recently.

But there are other explanations for the discrepancy. Perhaps saber-toothed cats lived in habitats where their remains were more likely to turn into fossils compared to where modern-cat ancestors lived. The likelihood that a dead animal will turn into a fossil depends a lot on conditions. In humid tropical rainforests, for example, carcasses tend to decay too rapidly for fossilization to occur. Probably for this reason, the fossil record for many modern-day tropical species is limited.

In addition, modern-day felines are for the most part very similar anatomically—it can be hard to tell them apart based just on their teeth and skeletons. As a result, paleontologists may be underestimating the number of species of conical-toothed cats that are contained in the fossil record. By contrast, the skeletons of saber-toothed cats are much more variable, making it easier to distinguish one species from the other.

A third possible factor is that studying saber-toothed cats is more exciting than studying fossils very similar to feline species alive today. As a result, the greater diversity of saber-toothed species may be a reflection of the greater paleontological attention they have received.

Testing these hypotheses is difficult, and all may be correct. Only more research will sort this out. Regardless, we're left with a dilemma: lacking much of a fossil record for conical-toothed felines, the ancestors of today's living cat species, how do we figure out what happened? Fortunately, evolutionary biologists have another trick up their sleeves, an approach for inferring how today's species evolved even in the absence of fossils. But before seeing how this is done, let's take a quick tour through the roster of contemporary cats.

JUST AS WITH THE FOSSILS, all of today's feline species are clearly cats; no one would look at even the most extreme species—the long-legged, small-headed cheetah—and not immediately recognize it as a cat. Perhaps it is for this reason that when we look at our household companions, we think big. I, myself, am guilty of this, but I have a lot of company.

My beloved Abyssinian was the runt of his litter. After his siblings had been sold, he was left alone all day in an empty apartment. Poor little dude—he was such a scaredy-cat when we first brought him home. But he grew into a magnificent, gorgeous, loving cinnamon cat. For his looks and big heart, we named him Leo, Latin for "lion."

Turns out that Leo is the sixth most popular name for male cats, one place behind Simba, which also means "lion," in this case in Swahili. In a similar vein, *Lion in the Living Room* was the title of both a Canadian Broadcast Company documentary and a popular book (both quite good). Clearly, people associate their feline friends with the lord of the African plains.

But there are other contenders for the housecat's alter ego. "Living-Room Leopards" was a fabulous article in *The New Yorker*. A classic book from 1920 bore the title *The Tiger in the House*, followed by *The Tribe of Tiger* seven decades later. Moreover, an analysis of names from a long-established pet cemetery found that the most popular name over the last 115 years has been Tiger.

So which species should we see in our mind's eye when we gaze upon our household companions lounging on the sofa—lion, tiger, or leopard? All three are magnificent animals, but there's an obvious disconnect between them and Smokey: the difference in size. Male African lions weigh as much as six hundred pounds; Siberian tigers (now rechristened as "Amur" tigers) are even heavier.* By contrast, a really,

*The largest feline ever was the South American saber-toothed cat, *Smilodon populator*, which tipped the scales at nearly half a ton—that's a big cat!

really big housecat weighs just a fraction of that, perhaps thirty pounds.*

Big cats are the celebrities of the feline world, the species that get all the attention, the stars of National Geographic Channel's Big Cat Week. But here's a little-known secret: of the forty-two species of wild felines,† the vast majority are about the size of a housecat.

Quick test: how many cat species can you think of that weigh less than fifty pounds? There's the ocelot, perhaps the most gorgeous of all cats, and the bobcat and its larger relative, the lynx (of which there are several species, some possibly exceeding fifty pounds). Can you name any more (hint: I've mentioned several in previous chapters)? Most people I've asked couldn't. Few have heard of the black-footed cat or the Borneo bay cat, much less the kodkod or oncilla. Clearly, the little-cat side of the feline family needs a better PR agent.

Color and spotting aside, all these little cats look pretty much like a housecat, except perhaps for the weaselly jaguarundi (a low-slung Central American species with a small, pointy head). There are some differences, of course, even beyond the beautiful spotted coats of many small species. Some are smaller than housecats—the tiny rusty-spotted cat weighs only three and a half pounds max—whereas the caracal, African golden cat, and serval can exceed thirty-five pounds.

Some have long legs (serval) or small ears (Pallas's cat). The fishing cat has webbed feet; the flat-headed cat's name says it all. Tails can be really long (the marbled cat) or short (the "bob" in bobcat). The ankle

---

**Guinness World Records*, which no longer uses weight as the measure of the largest animal of any kind, at one time put a very obese forty-seven-pound Australian cat in the top spot.

†There is disagreement on the exact number of cat species, revolving around cases in which experts disagree about whether two populations should be considered members of the same species.

joint of margays can rotate 180 degrees, allowing them to move headfirst down a tree, like a squirrel.

These outliers notwithstanding, when you take into account the range of cat species, domestic cats are much more similar in appearance to small wild feline species than they are to lions or other big cats. And don't be fooled by the romantic desire to cast your adventurous kitten as a tiger in the making; the differences between small and large cats extend well beyond looks and reflect significant differences in many aspects of feline life.

One major contrast is what they eat. Big cats tend to capture prey that are large relative to their size, sometimes exceeding their own body weight. By contrast, small cats tend to eat more diminutive prey that are only a small fraction of their own size such as insects, mice, and birds. Big cats also range over larger areas and reproduce more slowly than small cats. In all these respects, domestic cats are like their small-cat brethren.

The social behavior of domestic cats and lions is the major exception to this generality. I've already discussed their similarity as an example of convergent evolution, two species independently evolving the same trait. Another possibility, however, is that lions and domestic cats are close evolutionary relatives, in which case their behavioral similarities would likely have resulted from inheritance from a common ancestor rather than convergent evolution. Which is it?

Scientists have examined this question by studying the DNA of all living cat species. The methods are complicated, but for the most part, the more differences there are between two species in their DNA, the longer ago the species diverged from a common ancestor. Using these data, scientists have inferred the evolutionary tree of relationships among living feline species. Such depictions—"phylogeny" is the technical term—are like family genealogies. Closely related species occur near each other and can trace their ancestry to a recent common ancestor, just as brothers and sisters trace their lineage a short distance back to their parents. Distant relatives, like third cousins twice removed, occur on far-removed branches of the phylogeny; you have to work your way deeper into the tree, away from the tips—further back in evolutionary time—to find their most recent common ancestor.* How long ago species diverged can be inferred by assuming that differences in DNA between two species evolve at a roughly constant rate through time (with a lot of variation), what is sometimes called the "molecular clock."

The phylogeny reveals that the common ancestor of all of today's cat species lived about eleven million years ago. This ancestor then diverged into two lineages, the big cats (Pantherinae), with seven species, and the small cats (Felinae), with all the rest.

The common names for these groups aren't entirely accurate, however. All species in the Pantherinae are big cats (except for the mid-size, and stunningly beautiful, clouded leopard). But two members of the putatively little-cat Felinae, the mountain lion and the cheetah, are also quite large. The mountain lion is in many respects just a small cat scaled up to big-cat size, the reason that so many false reports of mountain lions turn out to be large tomcats seen at a distance. The long-legged cheetah, on the other hand, is the most distinctive cat species.

Domestic cats belong to the Felinae. Their closest relatives, members of a lineage that first appeared about seven million years ago,

---

*Because these studies are based only on living species, these ancestors are inferred to have existed, but they can't be assigned to specific fossil species. Scientists also build phylogenies that include fossils, but in those studies, the data usually come from comparisons of bones or other physical traits (except when DNA can be recovered from fossils, a topic we'll discuss shortly).

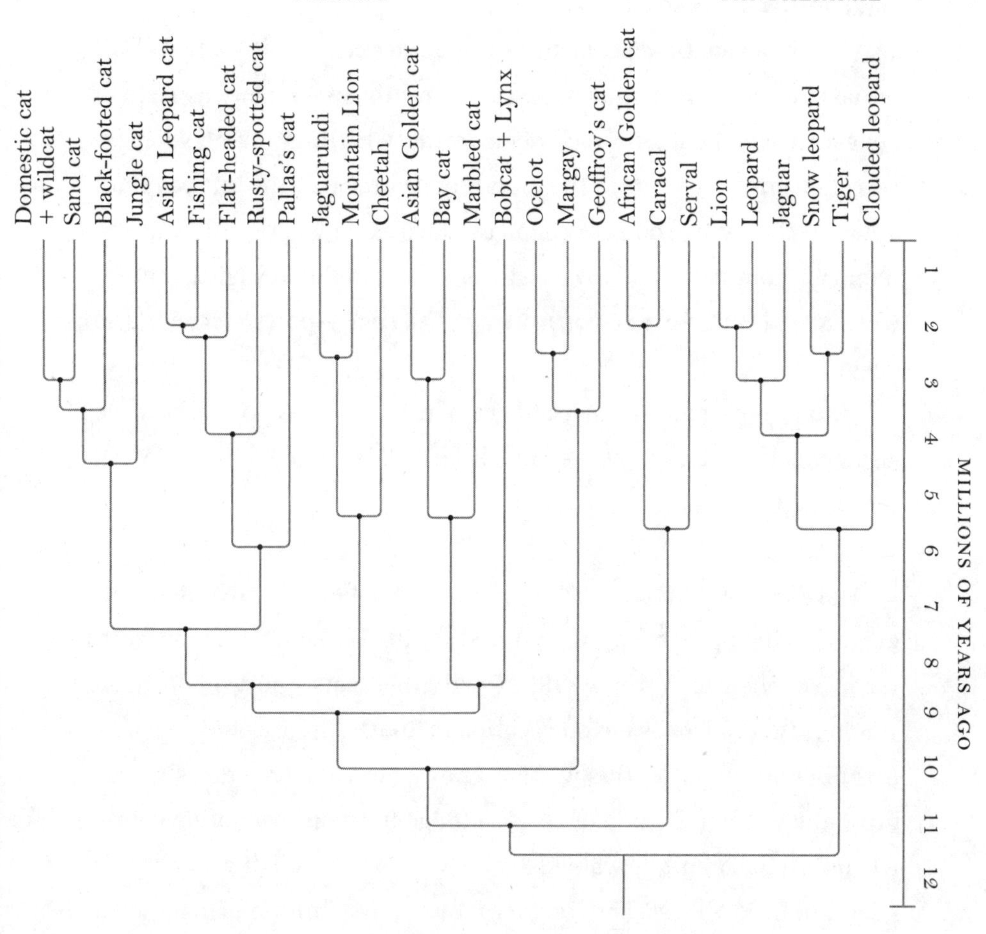

Evolutionary relationships (phylogeny) of living feline species. Species that share a common ancestor are more closely related to each other than to species not descended from that ancestor (ancestral species are indicated by the small black circles). To find when any two species share their most recent common ancestor, trace down the phylogeny from both species until their two lineages meet. Not all species are shown. For example, the group of Central and South American small cats includes not only the ocelot, margay, and Geoffroy's cat, but also the kodkod, two species of tigrina, and three others. Also, there are two species of clouded leopards.

are species similar in size and habits: the wildcats, sand cat, jungle cat, and black-footed cat.

The phylogenetic placement of the domestic cat with other small feline species corresponds to shared similarities in many aspects of appearance and habits. Social behavior notwithstanding, the domestic cat is more a margay in the manor or a caracal in the kitchen than a lion in the living room. Perhaps the next time you're thinking of a name for your cat, Geoffroy, Rusty, or Ozzie might be a better choice (referring, of course, to Geoffroy's cat, the rusty-spotted cat, and the ocelot).

With this as background, let's return to the question of how we know the domestic cat evolved from the African wildcat.

I'VE ALREADY SAID THAT African wildcats are nearly indistinguishable in appearance from domestic cats. But don't take my word for it. My mentor in the world of pedigreed cats, the woman from whom we got Nelson, went on vacation in South Africa a few years ago. A trained show judge, this woman knows her cats. One day she was browsing in the gift shop of a safari lodge in the middle of nowhere, when into the boutique walked a faintly striped, reddish-gray cat. My friend noticed the cat had long legs, but didn't think much about it. The cat wandered over and my friend began stroking him.

At that point, the woman behind the counter remarked that my friend was petting a South African wildcat! My friend took a closer look. Sure enough, the telltale sign, rusty-red color on the back of the ears: *Felis silvestris cafra*. That's how similar the two species are.

There are, of course, many domestic cats that you would never mistake for a wildcat. The kaleidoscope of colors, coat patterns, and hair lengths exhibited by *Felis catus* has no equivalent in the African wildcat. You won't find any long- or curly-haired wildcats on the African savanna, nor any black-and-white, orange, or white ones. Ditto for the physical characteristics that define many breeds.

The same is true for the African wildcat's close relative, the foul-tempered European wildcat. Stockier and darker in color with more prominent stripes than the African version, this wildcat is indistinguishable from some domestic cats. These markings, incidentally, are termed the "mackerel tabby" pattern because the striping along the back and down the sides sort of looks like a fish skeleton.*

A European wildcat.

That doesn't mean that it's impossible to tell a wildcat from a domestic cat, but doing so requires looking inside the cat. And if you do, there are two characteristics to examine, the length of the intestines and the size of the brain.

I was initially skeptical that domestic cats have longer intestines than their ancestors because most reports cited Darwin as their source. Don't get me wrong—I'm a big Darwin fan. Moreover, it doesn't

*Or so says the internet. I can find no definitive history of the term.

pay to bet against him, given that he has been proven correct about so many of his ideas—not just evolution by natural selection, but phenomena as diverse as the role earthworms play in aerating soil and how coral atolls form. Still, science was at a rudimentary level in the mid-nineteenth century, so I had doubts on this one, especially because Darwin's claim was based on research published in France in 1756. I'll be honest and admit that I did not go to the trouble of tracking down and translating Monsieur Daubenton's work, but my guess is that the Frenchman's finding that domestic cats have intestines a third longer than European wildcats was based on few actual data (I did check an 1896 follow-up study that presented measurements on a grand total of three cats).

I should never have doubted Darwin. A few years ago, scientists dissected specimens in a natural history museum and showed that feral domestic cats have intestines forty percent longer than European wildcats from the same area in central Germany; consistent findings subsequently were reported from Scotland (though, as far as I'm aware, no one has ever conducted a comparable study on African wildcats).

In theory, there's a ready explanation for this finding, one that Darwin himself proposed: "The increased length appears to be due to the domestic cat being less strictly carnivorous in its diet than any wild feline species; for instance, I have seen a French kitten eating vegetables as readily as meat." Meat is much easier to digest than plants or other types of food. For that reason, species that normally eat nothing but meat, like cats, have short intestines. Species with mixed diets—omnivores—have middling-length guts, and the plumbing of plant eaters is extensive.

Feral cats that scrounge a living around human habitations eat whatever they can find, including grains and other plants. It's easy to envision that in the early days of domestication, natural selection favored individuals with longer intestines to better digest the scraps they were eating, and that this led to the greater intestine length of today's domestic cats.*

Contrarily, in the brains department, domestic cats are less well endowed than their wildcat brethren. Two studies remarkably came to exactly the same conclusion that European wildcat brains are twenty-seven percent larger than those of domestic cats. A more recent study has confirmed that African wildcats also have more gray matter than domestic cats, though not quite as much as their European cousins. Reduced brain size is a common phenomenon in domesticated species and has been reported in sheep, pigs, horses, dogs, llamas, mink, and many other species. Differences in body size are statistically accounted for in these comparisons; the smaller brains are not just the result of overall smaller size.

Fear not, though, this does not mean that Sylvester and Fido (nor Porkie and Bessie) are stupider than their wild cousins. Rather, the reduction in brain size is concentrated in the parts of the brain associated with aggression, fear, and overall reactivity. This, of course, makes sense for domesticated animals living around people. High-strung animals prone to flight and stress wouldn't survive; natural selection would favor any individual less sensitive to such matters, and thus reduction in parts of the brain underlying these behaviors would be favored.

As far as anatomy goes, that's it: intestinal length, brain size, and, in African wildcats, rusty-red ears. If you're in the African bush or a European forest, those are the only traits that you can always rely on to determine which species you're examining.

---

*This does not justify feeding cats a vegan diet. Cats are finely adapted to a meat diet, so much so they're referred to as "hypercarnivores" by some researchers. A solely plant-based diet is not healthy for them.

. . .

Despite this great similarity, not all scientists agreed that the domestic cat descended from the wildcat. And considering that Africa, Asia, and South America are chock-full of small-cat species, there are a lot of other candidates. In theory, any of them could be the forebear of *Felis catus*.

We can, however, immediately eliminate the South American small cats. This group of nine species, which includes the ocelot and several other spotted beauties, has thirty-six chromosomes, whereas all other felines, including the domestic cat, have thirty-eight. This difference, and the fact that domestic cats are not known from ancient civilizations in the Western Hemisphere, would seem to rule out the South Americans.*

Still, there are a lot of domestic-cat-like little felines prowling the forests, plains, and deserts of Asia and Africa, and a number have been suggested as the domestic cat's ancestor. For example, the tiny sand cat is an adorable tawny desert dweller with an oversized head. Its candidacy was based on the hair covering the soles of its feet, just like those of Persian domestic cats.

Another suggestion was Pallas's cat, a YouTube favorite for its wizened-old-man-meets-Grumpy-Cat appearance and the small round ears on the side of the head. Its shaggy long hair provides warmth on the cold Asian steppes and evokes images of the hirsute Persian breed.

And then there's the beautiful spotted leopard cat. The many similarities it shares with Siamese cats—slender body, narrow head, long

---

*It's not impossible for a thirty-six-chromosomed ancestor to give rise to a thirty-eight-chromosomed descendant, but all else equal, an ancestor with thirty-eight chromosomes is more likely.

gestation period, friendliness of males toward kittens—suggested to some an ancestor-descendant relationship. Yet another contender was the jungle cat, which is quite similar to domestic cats, though longer of leg, shorter of tail, and a bit larger at twenty-seven pounds max. Plus, it was even kept as a pet by the ancient Egyptians.

These ideas were laid to rest in the 1970s by detailed studies of the skull and other aspects of anatomy. Working independently, German and Czech scientists concluded that none of these species was the ancestor; rather, skeletal similarities clearly indicated the strongest affinities are between domestic cats and wildcats. Subsequent DNA studies, which I've just summarized, then confirmed this conclusion, finding that the closest species to the domestic cat on the evolutionary tree is the wildcat.

These results were further bolstered in 2014 when a team of geneticists led by scientists at my own school, Washington University, sequenced the entire genome of a domestic cat named Cinnamon.* "Genome" refers to the genetic code of an individual, which in cats comprises more than two billion bits of DNA, termed "bases." Determining the identity of each base is obviously a gargantuan task; technological and computational advances in recent years, however, are making it possible for genomes to be readily studied. The scientists also sequenced, in somewhat less detail, the genomes of twenty-two other domestic cats, as well as two European and two North African wildcats.

A couple of quick terminological points: the working parts of the genome are the genes, of which cats have about twenty thousand (in the same ballpark as humans). Genes can be made up of hundreds of DNA

---

*So named for her brownish-orange fur, a trademark of the Abyssinian breed to which she belonged.

bases or millions. Alternative versions of a gene, differing in one or more bases, are called "alleles" or sometimes just "genetic variants."

The goal of sequencing and comparing these genomes was to identify genetic variants that were possessed by all domestic cats but not by any wildcats, thus revealing the genes that had evolved as the domestic cat diverged from its ancestors.

The scientists found that there are very few genetic differences that consistently distinguished domestic cats from wildcats: only thirteen genes showed evidence of having been changed by natural selection during the domestication process (of course, additional changes subsequently arose in some populations or breeds of cats).* By contrast, a similar study comparing dogs and wolves found almost three times as many genes involved in canine domestication.

This great genetic similarity between domestic cats and wildcats, paralleling the minimal differences in anatomy and behavior, seals the case that domestic cats evolved from the wildcat.

EVEN IF WE ACCEPT THAT domestic cats are descended from wildcats, however, that doesn't mean that domestication occurred in a single place. The wildcat has an enormous geographic range, encompassing much of Europe (except where they've been wiped out), Africa (except for the Congo rainforest and the Sahara Desert), and southwestern Asia. In theory, domestication could have occurred multiple times in different parts of that range.

Wildcats exhibit substantial anatomical variation across this ge-

---

*Technical point: genes can also diverge for random reasons. The scientists used complicated statistical methods to identify the changes resulting from natural selection.

ography. European wildcats are stocky with a broad head and a thick, dark coat. The African version is more svelte, with legs so long that when they sit on their haunches, they appear to be almost upright (think, not coincidentally, of classical Egyptian cat statues). Because of their legginess, these cats have a distinctive strut, with their shoulder blades protruding above their backs. Their faces are narrower and more angular than their European cousins, and their coats lighter and shorter. Asian wildcats are somewhat of a mixture, but in most respects more similar to African cats than to the Europeans.

These differences led to the idea that wildcats may have been domesticated in several places, the European wildcat giving rise to burly, broader-headed European cat breeds and the Asian wildcat begetting the shorter-haired, slimmer Asians. One line of evidence was presented by Czech researchers who claimed that the shape of the penis bone* differs between Persian, Siamese, and all other domestic cats. Based on these findings, the researchers proposed that domestication occurred separately in three different places.

This multiple-origins hypothesis is not unique to domestic cats. Dogs, cattle, goats, and chickens are all highly varied, suggesting the possibility that populations in different areas are the result of domestication from different wild ancestors. The other possibility, of course, is that these species were each domesticated a single time, and that today's great variety evolved subsequent to domestication.

How can we decide between these two alternatives?

*That's right—males of many species of mammals, but not humans, have a bone in their penis, technically termed a "baculum."

Six

Origin of Species

Enter Carlos Driscoll. Majoring in biology in the late 1990s, the University of Maryland, Baltimore County student was keenly interested in conservation. Driscoll wanted to save the world's species. But he didn't know which ones (though he did have a soft spot for snakes and lizards), nor did he know how he wanted to do it.

One day he was chatting with the chairman of the biology department, explaining that he was trying to figure out what to do after graduation. Turned out that an old college buddy of the chair's was doing cutting-edge conservation genetics research just down the road at, of all places, the National Cancer Institute (we'll get to why the NCI had a researcher on staff studying cat genetics in chapter fifteen).

Driscoll wasn't particularly interested in DNA research, but when you're facing an uncertain future, you don't turn down people offering to do you a favor. So he went to talk to Dr. Stephen O'Brien, who was, and still is, a leading figure in the field. The next thing you know, Driscoll was working in O'Brien's lab, studying the genetic variability of wild feline species. This is the kind of serendipity that often shapes scientific careers.

Driscoll worked in the NCI lab for six years, first as a paid technician, then as a master's degree student investigating genetic variation in cheetahs, lions, and pumas. By the time he was ready to start his PhD, Driscoll was hooked on DNA analysis. I've seen that happen with graduate students in my own lab. Initially intending to conduct fieldwork on ecology or behavior, the students are seduced into the lab by the immense power of genetic research. You go into the laboratory in the morning (or, for many grad students, you roll in some time around noon), work your butt off all day, and when you leave in the evening, you've generated a tremendous amount of new data. It may not be as exciting or romantic as leading a field expedition to some far-off, exotic locale, but you can make substantial progress very quickly. And the DNA can tell you a lot about how animals function, who's mated with whom, and how they've evolved through time.

Driscoll's challenge was to decide what species to study for his doctoral research and where to do it. Once more, fate intervened. Driscoll had a habit of volunteering to pick up visiting speakers at the airport, both because he's a nice guy and because, in his words, it "gave me an hour each way with a captive superstar of biology."

One such superstar was David Macdonald, an Oxford professor and one of the world's leading authorities on mammalian predators. The two hit it off.

Macdonald was interested in learning how to distinguish Scottish wildcats (the Scottish version of the European wildcat*) from domestic cats. We've already discussed how you can't reliably tell them apart by appearance. Even more difficult is distinguishing wildcats from the hybrid offspring produced by wildcat-domestic cat matings. Macdonald was hoping to find a genetic test to differentiate them, a necessary first step in trying to preserve the wildcat as a genetically distinct population. Driscoll by this time was already an expert in these methods. It was a match made in heaven. Off to Oxford went Driscoll for his doctoral studies.

As originally conceived, the project was narrowly focused on Scottish cats, but Driscoll quickly expanded its scope. Wildcats co-occur with domestic cats throughout their range, so "hybridization"—the technical term for interbreeding between two species or distinct populations—is potentially a widespread problem. And of course there was the riddle of domestic cat origins crying out for a sophisticated genetic investigation. If you're going to do your PhD on cat genetics, you might as well tackle the big questions!

Driscoll came up with a plan to obtain DNA from wildcats and domestic cats throughout the native range of the wildcat. It was an enormous undertaking and doing so from scratch would have taken many years, probably decades. Fortunately, many researchers in Europe had been studying their local wildcats, so samples were already available from numerous places.

Some of these researchers were already collaborators with the Oxford lab and were happy to contribute to the project. But many scientists at far-flung universities and government agencies around the globe had no connection to Oxford; they were doing their own thing,

*Once regarded as its own subspecies, the Scottish wildcat is no longer considered genetically distinct enough to merit separation from the European wildcat.

collecting data on local cats because it was a way of studying something close at hand. You need to know at this point that there is a pecking order of zoologists: the more charismatic the subject you study, the more prestigious. Studying lynxes and brown bears—very impressive. Jackals, somewhat less so. Feral cats? Not at all.

In other words, these cat researchers had a bit of an inferiority complex. Naturally, they were suspicious of an American they'd never met based at a hoity-toity English university asking them to turn over their samples. What was in it for them? How did they know they'd ever hear from him again, much less get credit for contributing to the project?

Driscoll had a plan to win them over. In what he describes as "the smartest thing I ever did," he'd shipped his fiery-orange BMW motorcycle to England when he moved to Oxford. When it came time to visit the researchers and request their help, he didn't book a plane flight and stay at a nice hotel like a big-shot researcher. Rather, he hopped on his bike, crossed the Channel by hovercraft (preferably) or Chunnel (if need be), and motored his way through Europe to visit researchers in Hungary, Bulgaria, Slovenia, Croatia, Serbia, Montenegro, and other destinations. His natural man-of-the-people approach worked like a charm; rather than awkward, stiff encounters, he ended up with invitations to stay overnight and enjoy home-cooked meals. And he came away with the samples.

Africa and Asia were a different story. Only in South Africa were there researchers already collecting samples. If Driscoll wanted to sample the cats in most of these areas, he'd have to catch them himself.

Remember, Driscoll had spent his scientific career to this point analyzing DNA in the lab. Despite tagging along on occasional lab research trips, his field experience was limited. By his own admission, "I was pretty bad at catching cats at the beginning." But as the project progressed, he got better. A lot of the success came from learning exactly where to place the trap—either a rubber-padded leghold or a Havahart walk-in—and how to lure the cat to step into it.

Once he learned the fine art of trap placement (a feather dangling nearby was a seemingly irresistible lure), the only issue was that occasionally he caught other types of animals, some of which were a bit dicey to release, including honey badgers (you know, the ones that "just don't care"), skunks, and monitor lizards. The most memorable misfire was the Havahart trap that caught a mother warthog and several of her piglets. By the time Driscoll returned to liberate them, the trap was completely destroyed, but somehow the door had remained latched (the warties, however, were fine and took off, tails held high, as soon as the door was pried open).

Driscoll traveled to Asia and Africa in quest of cat samples, visiting Israel, Azerbaijan, Kazakhstan, Mongolia, China, Namibia, and South Africa. Even though much of this work was conducted shortly after 9/11, he never had a problem getting around. Geopolitics did impact the scope of the study, however. Egypt plays an important role in the cat story, but research visits there and to much of the rest of the Arab world became difficult. A scheduled trip to Iran—the result of an extensive, agonizing planning process—was canceled the day after President Bush's "Axis of Evil" speech in 2002. Fortunately, the African wildcat is widely distributed in the Middle East: Driscoll was able to collect samples from Israel, as well as accessing several from Bahrain and the United Arab Emirates already in the collection of one of his collaborators.

Ultimately, Driscoll ended up with an extraordinary collection of 979 samples of wildcats from throughout their range and domestic cats from around the world.

The samples were varied. Most were blood samples collected by scientists during routine monitoring. But Driscoll didn't pass up other opportunities, such as taking tissue samples from road-killed cats he came upon. He also snipped small bits from taxidermy mounts in natural history museums. A bit of the sphincter muscle—aka, the butthole—was particularly useful because curators didn't mind losing a sliver of that as opposed to, say, the tip of an ear.

Undoubtedly, the most unusual source was an ankle-length coat owned by an eagle hunter in western Mongolia. The Kazakh people of that region train golden eagles to catch Pallas's cats and foxes, which they use for fur. The ankle-length coat Driscoll saw (and tried on) was made from forty Pallas's cats and lined with Asian wildcat fur; by clipping a small piece of each hide, he not only got samples from the local wildcats, but hit the jackpot for the uncommon Pallas's cat (which he could use in subsequent studies).

Back in the lab, Driscoll processed the samples, extracting the DNA and analyzing it. This was before the days when entire genomes could be easily and cheaply sequenced, so he focused on several specific genes. The result was thousands of bases from each sample. By comparing these DNA pieces, he could infer how closely related individuals were.

Some of the results were as expected. Wildcats in different parts of the world are, indeed, genetically distinct from each other. But there was a surprise: there are four genetic groups of wildcats, not three. Not only are European and Asian wildcats distinctive, but the African wildcat is actually two genetically distinct groups, the South African and North African wildcats (the latter including cats from Turkey, Israel, Saudi Arabia, and nearby countries, and hence sometimes called the Near Eastern wildcat). Based on the extent of genetic differences, Driscoll estimated that the four groups had been genetically isolated from each other—that is, not interbreeding and exchanging genetic material—for well over one hundred thousand years (more recent studies have suggested that this is a substantial underestimate).

And there was yet another surprise. One of the most poorly known feline species is the Chinese mountain cat, which lives at high elevations on the Tibetan Plateau (the area known as the "Roof of the World," near Mount Everest). So little is known about this species that one expert wrote that it has short legs, another that it has long legs. Everyone agrees, though, that it has the stocky physique and long hair of the Scottish wildcat, but a lighter coat more similar to non-European wildcats. Although some had suggested that it was a wildcat, most authorities considered it a separate species.

Driscoll's work put the kibosh on that idea. Using the DNA data, Driscoll built an evolutionary tree of relationships among the different types of wildcats and related species. As expected, the wildcats are all more closely related to each other than any of them is to the sand cat, which previous studies had shown to be closely related to wildcats. But the phylogenetic placement of the Chinese mountain cat was

A Chinese mountain cat.

a surprise (at least to most), right in the middle of the wildcats, the closest relative of the Asian wildcat. In other words, the Chinese mountain cat is, indeed, a fifth type of wildcat.

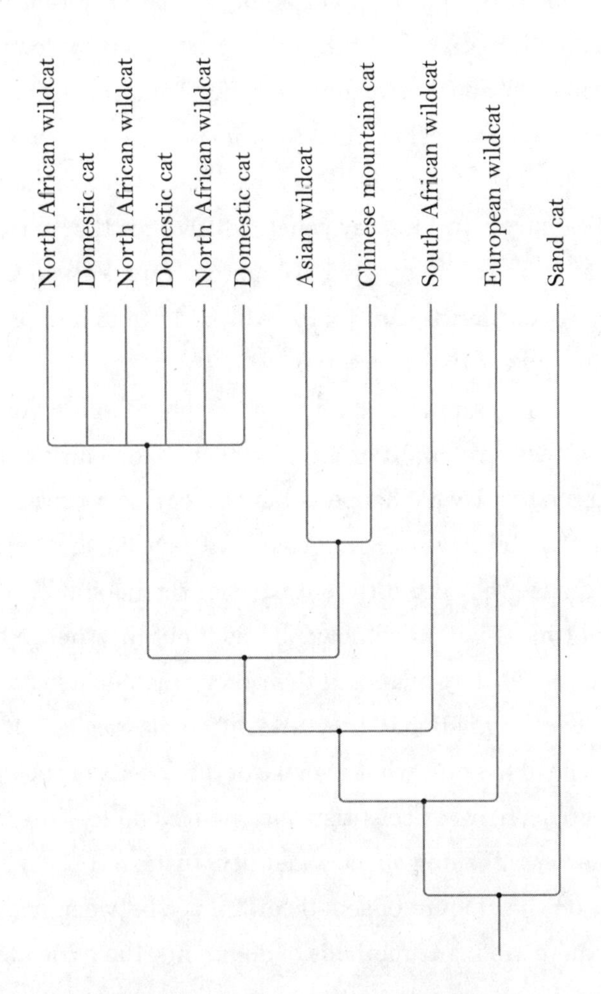

Evolutionary relationships of wildcats and domestic cats. The representation of the North African wildcat and domestic cats is a simplification to emphasize that the two do not form separate groups; rather, they are intermixed. The lengths of the lines are not meant to be proportional to time elapsed, in contrast to the phylogeny on page 73.

. . .

IF YOU REMEMBER your high school biology class, you may recall that two populations are considered different species if their members are unable or unwilling to mate with each other, or if they do mate, they are incapable of producing fertile offspring. The rationale is simple. If species can't exchange genes, then they are on their own evolutionary trajectory. But if they can exchange genes, then they are not independent evolutionary entities—any genetic difference that arises in one can be passed on to the other. This approach works well if the two populations occur in the same place. All you have to do is go out and study their reproductive interactions.

But how can we know if members of two populations would interbreed when they live on different continents? You can put them together in a zoo to see what happens, but that can be misleading. Many species that would have nothing to do with each other in nature—perhaps because they use different parts of the habitat or are active at different times—will nonetheless breed if put together in the same cage. Of course, if they mate and their offspring don't survive or are sterile, you've learned that they are different species. Otherwise, though, it's hard to know what to make of the results of such projects.

For this reason, many scientists have soured on the interbreeding criterion for establishing species identity. Instead, they make their judgments on the amount of genetic difference between populations. Although there are many methods of doing this, the basic idea is that populations that are greatly different genetically must have been evolving with little genetic interchange for long periods of time and thus should be accorded species status. Just how different the populations have to be is subjective and a matter of debate.

This new approach is the reason that the number of recognized feline species has increased markedly in recent years. Genetic studies have shown that in many cases, geographically isolated populations of what used to be thought of as one species are genetically greatly different from each other and so now are considered to be two or more species. That's the reason we now recognize two species of clouded leopards, one from mainland Asia, the other from Indonesia. Similar splitting of species has occurred in leopard cats, tigrinas, and others.

The two approaches aren't as different as they may seem. As a general rule, the more genetically different two populations are, the less likely they are to be able to interbreed. In many cases, the two methods produce the same result.

In the case of the five populations of wildcats, we don't have evidence of interbreeding; all we have is the genetic data. Three books on the diversity of feline species published in the last ten years, all by experts in cat biology and taxonomy, come to three different conclusions. In one, all wildcats—including the Chinese mountain cat—are considered one species, with five subspecies. Another goes old-school and recognizes the Chinese mountain cat as a different species, lumping the other four as subspecies of the wildcat. The third also recognizes the Chinese mountain cat as distinct, but raises the European wildcat to species level, presumably because it is anatomically distinctive from the other three, which are left as subspecies of an African-Asian wildcat species.

There's no objective way to decide among these three taxonomies. My preference lies with the first option, considering them all members of one wildcat species, *Felis silvestris*. For our purposes, it doesn't really matter. Species or subspecies, the five are genetically distinct from each other.

. . .

But let's get back to domestic cats, the central focus of Driscoll's work. If domestication had occurred multiple times from different ancestors, then we would expect one of two possible outcomes. One possibility is that domestic cats from different parts of the world would group with the corresponding wildcats from that region: Asian domestic cats and Asian wildcats would be nearest relatives on the phylogeny, British housecats with Scottish wildcats, and so on. The genetic signature of multiple domestication events would be clear.

Alternatively, even if the domestic cat had been domesticated multiple times from different wildcat ancestors, it is possible that the genetic differences among these populations were subsequently homogenized by movements of cats from one place to another. In that case, the domestic cat gene pool* would be a mishmash of DNA from all wildcat populations, perhaps causing domestics to form their own genetic cluster distinct from all the wildcat subspecies.

Driscoll's results disproved both of these possibilities. The DNA analysis placed domestic cats from all over the world squarely with the North African wildcat. In fact, the two are so similar that it was not possible to distinguish the two groups—they were intermixed genetically. No wonder the study on cat genomes a few years later found so few genetic differences characterizing domestic cats compared to wildcats.

Driscoll's study closed the door on the idea that cats have been domesticated multiple times from different ancestors. The domestic cat

---

*"Gene pool" refers to all the genetic variation—all the alleles for every gene—present within a population.

originated from the North African wildcat, and only from the North African wildcat.

Still, some questions remained unanswered. In particular, was the North African wildcat domesticated once, in one place at one point in time, or in multiple places? The species is apparently predisposed to associate with humans, so it's easy to envision North African wildcats living among people and being domesticated repeatedly in many places throughout the subspecies' geographic range.

The DNA data favor this latter possibility. If domestic cats originated in one place, they would likely have a limited amount of genetic variation, yet Driscoll's study found exactly the opposite—domestic cats today are extremely variable genetically. Most likely, Driscoll contends, this diversity is a result of North African wildcats being repeatedly domesticated at multiple places and thus containing genetic variation from many different populations.

One important implication of Driscoll's finding is that to understand domestication of the cat, we need to look specifically at the North African wildcat. Some important studies, such as Nicholas Nicastro's study on vocalizations, were conducted instead on the South African wildcat (of course, at the time of Nicastro's study, we were not aware of the genetic distinctiveness of North and South African wildcats). In other cases, we'll need to dig deep to figure out whether reports on wildcats in the wild or reared as kittens refer to members of the North or South African subspecies. The deeper question is the extent to which differences in behavior and anatomy exist between the two subspecies; as far as I know, no data currently exist on this point. The fact that they were lumped together as the African wildcat, however, highlights their overall similarity.

As for the original goal of the project—to develop a genetic test to identify hybridization between domestic cats and wildcats—the study was a complete success. Domestic cats are derived from North African wildcats, and North African wildcats are genetically distinct from other wildcat subspecies. So if you're examining the genes of a putative Scottish wildcat and you find some North African wildcat DNA, you know that you don't have a cat that is genetically one hundred percent Scottish wildcat, but rather one that has a domestic cat as one of its ancestors (the alternative possibility is that your putative Scotscat has a North African wildcat as an ancestor, but then you'd have to explain how a North African wildcat got to Scotland).

Indeed, Driscoll found evidence of such hybrids in all wildcat populations. In some cases—such as in Kazakhstan and Mongolia, as well as various places in Europe—the majority of wildcats carry some North African wildcat heritage in their genetic makeup. Wildcats the world over appear to be interbreeding willy-nilly with domestic cats.

In the subsequent fifteen years, a number of researchers extended Driscoll's approach and developed ever more sensitive tests for genetic mixing of domestic cats and wildcats. Their findings are in agreement: some degree of hybridization occurs in almost all wildcat populations. Conversely, wildcat DNA finds its way into domestic cat populations as well. Chinese researchers, for example, were surprised to find Chinese mountain cat DNA in some local domestic cats.

HYBRIDIZATION IS A MAJOR concern to conservationists. Their goal is to prevent species from going extinct.* We usually think of threats that kill off individuals of a species: overhunting, destroying their habitat, wiping out their food source, and so on. But another threat is genetic pollution. If enough genetic material from another species is

*Also subspecies, but for simplicity, I only refer to species.

introduced, then what is it you're preserving? Certainly not the original species that had evolved over thousands or millions of years.

Consider the Scottish wildcat. Sporting black stripes on a gray background, the "Tiger of the Highlands" deserves its nickname. Brawny, with a big head, thick coat, and a bushy, ringed tail, it's easy to envision these cats moving through the snow like miniature versions of their Siberian namesake. Formerly found throughout Great Britain, these cats are now restricted to mountainous areas of northern Scotland due to habitat destruction and persecution by game wardens.

In recent years, forests have regrown in many places, and the cat is now legally protected. Yet, a new threat has appeared. In Britain, as in the United States and elsewhere, domestic cats are all over the landscape. And when they run into wildcats, romance often ensues. Because there are so many domestic cats—both owned and unowned—and not so many wildcats, a large proportion of wildcat couplings are with domestic cats. Conservationists estimate that at most there are only several hundred Scottish wildcats remaining in the wild. That's not to say that there aren't any cats out in the Scottish Highlands. Quite the contrary, there are many. It's just that most of them have domestic cat ancestry in their family tree and so are considered to be hybrids rather than Scottish wildcats.

We have to ask, though, how much of a problem is this? Many of these hybrid cats are virtually indistinguishable from their genetically unmixed counterparts. Scientists have strived mightily to identify external characteristics that always differentiate wildcats from hybrids, but even the best external indicators of Scottish wildcatishness—the presence of four stripes on the nape of the neck (instead of only two), two stripes on the shoulder, unbroken vertical stripes on the side, and a tail tip that is solidly black and blunt, rather than tapering to the end—are not entirely reliable.

Given this indistinguishability, does it really matter that hybridization is occurring as long as there are wildcat-looking cats out in the woods, filling the wildcat niche in the ecosystem? Indeed, scientists have come to realize that occasional interbreeding between animal species of all sorts is more common than previously realized, even in natural settings little disturbed by humans. Despite the definition of species as groups that don't interbreed and produce fertile offspring, it now appears that the definition is more of a guideline than a hard-and-fast rule. Two species can, at least sometimes, maintain their evolutionary independence despite exchanging genes with each other. So maybe hybridization is not always such a bad thing.

On the other hand, some conservationists have a philosophical distaste for hybridization, a desire to maintain the genetics of species the way they were before people came along. Such a view seems antiquated in view of the modern understanding that hybridization sometimes does occur naturally between species, as well as with regard to the realization that humans have altered so much of the world that it's too late to try to preserve nature the way it was, rather than dealing with the reality of the way it now is. In addition, most scientists consider the North African and European wildcats to be members of the same species. So hybridization with domestic cats is just bringing in genetic variation from another subspecies of the same species. Perhaps that's not a big deal.

But there's a deeper concern that the consequences of hybridization may not be so harmless. If hybridization runs rampant, we could end up with a cat population roaming the Scottish moorlands that consists of individuals looking more like the ones in the alley behind your house than a classic wildcat.

Indeed, already this is somewhat the case. Although most cats on the Scottish landscape look more or less like wildcats, not all of them do. About one cat in six is black, a color never seen in Scottish wildcats until recently. Research reveals that these animals almost surely got their black coloration from a domestic cat ancestor.* Another fifteen percent sport a variety of colors, including cats that are black and white, gray and white, pure white, orange, some with swirly patterns, and some even have long hair. The concern, of course, is that if hybridization continues unabated, the distinctive appearance that is the essence of the Scottish wildcat will disappear.

And it's not just about appearance. Remember the differences in intestine length and brain size between European wildcats and domestic cats? Hybrid cats often have intermediate characteristics, and thus their digestive efficiency is likely reduced and their responsiveness to external stimuli probably muted.

So perhaps there is a reason to be concerned about hybridization between wildcats and domestic cats. A hybrid swarm of wild cats may be quite different from the native wildcats that used to live in Scotland: they may behave differently, survive differently, and affect the ecosystem differently. It might be a Scottish environment with cats, but possibly one very different from the one that existed before people and domestic cats arrived on the scene.

Yet, on the third hand, one could argue that if big brains, short intestines, and tabby stripes are beneficial for survival in Scottish moors, then natural selection will work its wonders and the population will retain its essential wildcat traits: as quickly as maladaptive traits are introduced by hybridization with domestic cats, natural selection will weed them out. For example, there's probably a reason that

---

*An interesting parallel: some wolves in Yellowstone and elsewhere in North America are black, a color never previously seen in wolves. Genetic analyses indicate that this color is the result of interbreeding with dogs sometime, somewhere, in the past.

black-and-white cats don't occur naturally in any cat species. Such cats probably are more apparent to both predators and prey, and thus have lower survival rates. Similarly, if big brains are important for survival in the Scottish wilds, then natural selection would be expected to favor the biggest-brained cats. Scottish cats would remain Scottish wildcat-like even with the influx of domestic cat DNA.

So it's hard to know what the end result of hybridization will be. Scottish conservationists are now taking reasonable steps to minimize hybridization by encouraging pet owners to keep their cats inside and to neuter them to minimize mating with wildcats. In addition, they are trying to reduce the number of feral domestic cats on the landscape through a variety of approaches. This strategy makes sense.

At the same time, obsessing over whether cats have any domestic blood in them seems like a lost cause. If a cat looks like a Scottish wildcat, let it be. The best course is to focus on limiting the influx of new feral domestic cats as well as to remove cats (hopefully, finding a good home for them) that, based on their appearance, are obviously domestic cats or hybrid offspring.

DRISCOLL'S FINDINGS on the pervasiveness of domestic cat–wildcat hybridization has important implications that extend beyond Scottish wildcats. The fact that wildcats and domestic cats readily mate in the wild and produce fully fertile offspring means that by the standard criterion, domestic cats and wildcats belong to the same species. Indeed, for this reason, some scientists classify both as *Felis silvestris*, with the domestic cat comprising the subspecies *Felis silvestris catus*.

Other scientists, including me, don't follow this convention and instead refer to the domestic cat as *Felis catus*. Many domesticated animals have different scientific names than their wild ancestors, even though they can interbreed—for example, dogs and wolves (*Canis domesticus* versus *Canis lupus*). We use these different names not to indicate that domestic and wild species are reproductively incompatible—they usually aren't. Rather, these different names emphasize the great anatomical and behavioral divergence of the domesticated species from its ancestor (greater in most domesticated species than in cats); treating domestics as distinct species also highlights the role humans played in the evolutionary changes that occurred during domestication.

You say toe-may-toe, I say toe-mah-toe. Whether we call domestic cats *Felis catus* or *Felis silvestris catus* doesn't really matter—the scientific reality is that the domestic cat and the wildcat are members of the same biological species: they interbreed readily and their hybrid offspring can be hard to distinguish from non-hybrid members of either species. This lack of differentiation highlights how little the domestication process has moved cats from their wildcat roots.

The existence of extensive interbreeding between domestic cats and wildcats adds one further wrinkle: it makes it very hard to study the evolution of the domestic cat. The reason is that in order to figure out how domestic cats have changed from their ancestors, we need to know what their ancestors were like. In an ideal situation, we would have fossils of the ancestors from the time when domestic cats started to diverge, a before-and-after comparison of ancestor and descendant.

But we don't have such fossils, so instead we compare domestic cats to modern African wildcats with the assumption that the African wildcat today is the same as the ancestral African wildcat from which domestic cats descended. Because of hybridization, this assumption may not be correct.

The problem is that when domestic cats breed with wildcats today, any trait that domestic cats evolved can be transmitted back to the wildcats. Suppose, for example, that the ancestral African wildcat had purple polka dots instead of stripes. Then, when domestic cats evolved from African wildcats, for some reason they evolved to replace the polka dots with stripes (perhaps stripes provided camouflage useful for hunting around human habitations, or maybe people just preferred stripey cats). At that point, the two types of cats would be readily distinguishable. But then, because the two interbreed, the allele for stripes might pass from the domestic cat to the African wildcat. If stripes were also beneficial out in the bush, natural selection would favor that allele and wildcats would evolve to change from polka-dotted to striped.

In other words, the interbreeding between these two species tends to homogenize their gene pools, eliminating differences between them. As a result, one reason that domestic cats may seem to have evolved very little from their wildcat ancestors is not that they haven't evolved, but rather that all their evolutionary advances subsequently were gained by today's wildcats as well.

For example, the conclusion of the genome comparisons was that very few genes changed during cat domestication. But an alternative possibility is that initially domestic cats evolved differences in many more genes, but that the domestic cat version of many of these genes was passed back into wildcat populations via hybridization.

Similarly, these considerations suggest that the amiability of modern-day African wildcats may be the result of hybridization with domestic cats, so perhaps the conviviality of our household companions really was a large evolutionary leap, one that housecats have passed back to their ancestors. Ditto for the suggestion that African wildcats have group-living tendencies. Remember the colonies of African wildcats living in fennec fox holes? Instead of indicating a latent tendency for sociality in African wildcats, an alternative possibility is that the British explorer didn't realize he was observing hybrids between

African wildcats and domestic cats, in which case the group living he saw may have been a result of the colony members' domestic cat heritage.

One way to discount this possibility would be to know that hybridization hasn't occurred in a wildcat population under study. However, unless one samples a population that occurs in a place where there are no domestic cats (good luck on finding that), definitively knowing that hybridization hasn't occurred is very difficult.

There is another option, however, with regard to genetic differences. If we could find fossils of wildcats from pre-domestic-cat days and extract DNA from them, we could see what their genomes were like before they came into contact with domestic cats. That sounds like science fiction, but as we'll see in chapter eight, it's not as impossible as it sounds.

BUT ENOUGH ABOUT HYBRIDS. Let's get back to the bigger question of cat origins. Although Driscoll's work demonstrates unequivocally that the domestic cat descended from the North African wildcat, it doesn't tell us how domestication happened, much less exactly where and when. A number of different scenarios have been proposed to explain the origin of *Felis catus*. Evaluating those ideas requires moving from the field of genetics to the study of ancient civilizations. Let's consider what these ideas are, and what the archaeological data have to say about them.

Seven

Digging Cats

The earliest evidence of cat-human association, dating to about ninety-five hundred years ago, comes from two graves on the island of Cyprus. One of the graves contains a human. Sixteen inches from the person's feet was a second, smaller grave containing an eight-month-old cat, carefully laid on its side and well preserved. The person was buried with prized objects including axes, polished stones, and ochre, which suggests that the cat was a treasured possession as well. The large size of the cat also suggests that he was well fed, further evidence that he may have been a tame household animal, perhaps even a beloved pet.

Across the many species that have been domesticated—from ducks to dogs to donkeys—two general paths have been taken. In one route, humans took charge from the outset, controlling the animals and, sooner or later, directing their breeding. In some cases, this occurred gradually, involving species that humans originally hunted, but through time started to manage in herds by limiting their migrations, enclosing them in large areas, and taking care to avoid killing females so that the herd would grow more quickly. Eventually, people began to direct the reproductive process, choosing which individuals could reproduce based on the possession of desired characteristics. The domestication of many common barnyard species such as cattle, goats, and sheep probably began this way.

Obviously, this managed population route doesn't apply to cats. Herding cats? The thought is laughable, and there isn't any reason to suspect that humans captured wildcats and started selectively breeding them.

In the other route to domestication, the animals took the lead in adapting to living with us. As human civilization emerged, our habitations provided opportunities in the form of food, shelter, and safety from predators. Many species took advantage and became "human commensals," living in our midst and benefiting from our largesse, some to our great displeasure (think rats and cockroaches).

In some cases, this association continued into domestication as the animals slowly began to depend more deeply on humans. They adapted to live in close association with us; in turn, we more intentionally provided resources and, in many cases, began making decisions about which individuals could reproduce.

Wolves, for example, may have originally been attracted to scavenging in refuse pits near human settlements. The boldest, least fearful members of the pack may have had the advantage because they would be spooked less readily and thus would get more food. In turn, kindhearted humans may have started throwing scraps to the wolves, further favoring the friendliest—or at least most human tolerant—members of the pack.

Perhaps wolves began to see the village as their territory and started defending it against other animals by vocalizing when someone or something came near. Humans may have begun seeing value in having the wolves around and so treated them even more nicely. The two species became closer. Generation after generation, those wolves least afraid of humans would have benefited the most and would have had more offspring as a result. In due course, the wolf transformed into the dog. Once they were living among us, we eventually started choosing which dogs could reproduce; dog breeds sculpted to serve different purposes were the outcome.

Most researchers paint a similar picture for the domestication of cats. About ten thousand years ago, agriculture began. People transitioned from living in a hunter-gatherer lifestyle—constantly on the move looking for plant and animal food—and settled down to become farmers. This was the dawn of the Neolithic revolution, and it occurred first in an area known as the Fertile Crescent, encompassing what is today parts of Iran, Israel, Syria, Turkey, and other countries.

One of the advantages of an agricultural lifestyle is that when conditions are good, food can be grown in ample quantities and stored for leaner times. But in nature, opportunities rarely go unexploited. The archaeological record reveals that rodents of various sorts quickly moved in to take advantage of the bountiful food source sitting in storage.

The Fertile Crescent lies within the natural geographic distribution of the North African wildcat. And just as rodents exploited the newfound availability of seeds and grains, wildcats took advantage of the new abundance of one of their favorite menu items.

As you probably know, cats and other household animals vary in temperament. Some are curious and bold, others timid and fearful. In recent years, a growing cadre of behavioral scientists have focused on the differences in personality among not only domesticated animals, but also wild species. And not surprisingly, they've found that personality differences exist in most species. In my own laboratory, for example, a postdoctoral researcher demonstrated that lizards differ in their behavioral tendencies and that such differences affect their survival in the presence of predators.

Imagine, then, what happened when the first human settlements emerged. Suddenly, there were clusters of people living in an area. Some wildcats, the more cautious ones, wouldn't have anything to do with the villages and stayed away. But more adventurous or curious cats might check them out, even hang around. These cats would have been rewarded by the extra food they discovered in the form of plentiful rodents; perhaps scavenging from human trash piles also added to their diet. In addition, because larger predators probably avoided these settlements, wildcats may have been safer living there than elsewhere (though the presence of already domesticated dogs may have counterbalanced that benefit).

More food and fewer predators translate into longer lives and more offspring. Natural selection would have favored the evolution of cats that were not afraid of people and that were attracted to living among us. Presumably, humans would have tolerated the cats, perhaps even encouraging their presence for their pest-control services. The cats, in turn, would have evolved not only to associate with humans, but also to adapt to new circumstances. In particular, the abundant food would have led to the presence of many cats. Notoriously solitary in their former settings, settlement cats would have needed to change their ways to suppress their anti-social tendencies.

At this point, cats would have become commensals, adapted to living around humans, in a manner like that proposed for wolves. A similar process may be going on today with raccoons, foxes, opossums, and other species common in urban settings.

But commensal and domesticated species are not the same thing—think about house mice and house sparrows, so well adapted to living around us but hardly domesticated. Early cats were probably similar to those that lived in remote Sudanese villages early in the twentieth century, about which a local official was "very amused at the idea of anyone owning a cat: he said that cats live about villages and enter houses at night, but that they are wild animals and do not let people touch them."

So how did cats go from household hangers-on to treasured pets? It's not hard to imagine what happened next. Perhaps people tried to attract cats by providing food and shelter so as to benefit from their rodent-catching skills. In turn, the friendliest cats got extra food and people started valuing having them around just because they enjoyed their presence. Natural selection favored the cats that interacted most effectively with people. Presumably, this led cats to be better taken care of, and thus to live longer lives and have more kittens. Any genetic mutation that made a cat friendlier to humans would be favored and would spread through the population. Before long, *Felis silvestris lybica* transformed into *Felis catus.*

Some researchers have cast doubt on this scenario by disparaging the domestic cat's rodenticidal tendencies. Carlos Driscoll himself wrote that "cats do not perform directed tasks and their actual utility is debatable, even as mousers. In this latter role, terrier dogs and the ferret (a domesticated polecat) are more suitable. Accordingly, there is little reason to believe an early agricultural community would have actively sought out and selected the wildcat as a house pet."

This seems like a pretty bold claim, given the housecat's well-deserved reputation as a depredator of wild species. What's the evidence?

Several studies indicate that, for all their ferocity, cats are not effective at regulating rat populations. A study on farms in England conducted in the 1940s demonstrated that if rats were eliminated, the presence of cats usually would keep them from reestablishing in the farm buildings. However, agricultural fields a short distance away remained full of rats.

In a study in downtown Baltimore, a researcher observed rats at night while sitting in a car parked in various alleyways. During this time, "although cats and large rats were frequently observed in close proximity . . . generally these species coexist peacefully," the researcher wrote, adding, "I witnessed adult cats pursuing large rats on only five occasions and pursuits always terminated prior to physical contact."

More recently, researchers studying rats at an industrial waste recycling plant in Brooklyn made similar observations. The researchers were studying the rat population at the plant when five feral cats showed up. Concerned that the cats would ruin their study by decimating the rat population, the researchers made the best of a bad situation and set up motion-triggered video cameras to document the onslaught. To their surprise, the cats were pretty ineffectual. Over the course of seventy-nine days, the cameras recorded 259 videos in which cats appeared, but stalking was observed only twenty times. The only two kills occurred when a cat successfully pulled a rat from a hiding spot. Moreover, "the lone predation attempt on the open floor was a failure when the rat stopped running, and the cat also ceased the chase, only to stare at it."

So Driscoll may have a point—these studies suggest that cats may not be talented predators on large rats in urban and farm settings.

On the other hand, the scientific literature is replete with reports of cats eating rats in more natural settings. Indeed, rats are often the most common prey item of island cats. One study of feral cats in New Zealand, for example, found that rats were the staple prey item, accounting for nearly half by weight of all the food eaten; in another study, also from New Zealand, ninety-three percent of cats had eaten rats (as revealed by picking through their poops). This penchant for rativory is not a recent one: an archaeological dig from Egypt discovered the remains of a large cat with five adult rats in his stomach. Moreover, in recent years, animal welfare groups have established programs such as Chicago's Cats at Work, which adopts out cats to businesses and residences with a rat problem.

This debate may miss the point, however, for a simple reason: rats aren't the only problematic species in agricultural areas; mice and other small rodents can be a huge problem as well. Even if the rat-killing capacity of cats is overstated, no one can deny their effectiveness in mouse control. In addition, in some places today, cats are prized for their predatory attention to snakes. Tomb paintings suggest that was the case in ancient Egypt as well.

Bottom line: the traditional hypothesis that cats were valued in ancient times for their ability to keep pests in line remains very plausible.

But there are alternative possibilities. For example, perhaps somewhere people started capturing litters of kittens, raising them, and then breeding the friendliest ones. North African wildcats are somewhat friendly already;* perhaps it wouldn't have taken too many generations of selection to produce the friendly cats we know and love. Call this the "selection-to-create-a-pet" hypothesis. A very similar idea has been proposed for the domestication of the dog, though it seems to have fallen out of favor.

A variant of this possibility is that cats were domesticated for religious reasons. Perhaps cats became revered, prompting priests to initiate a breeding program on temple grounds to produce docile cats for their ceremonies.

So there are a number of hypotheses about how cats became domesticated, but what about where and when?

THE CLASSICAL APPROACH to investigating what happened in the past is to directly examine the historical record. Archaeology is the discipline that studies past human civilizations, and the study of human-animal interactions in the past has its own subdiscipline, zooarchaeology (not the excavation of ancient zoos, as I thought when I first heard the term at age fifteen, but the study of animal remains in human archaeological contexts).

What we can say for sure is that by about thirty-five hundred years ago, *Felis catus* existed as a household pet in ancient Egypt. We know that from depictions of cats as family members—under the dinner table, on boating trips to the local marsh—in paintings on tomb walls. Images of cats are almost entirely absent from Egyptian iconography before four thousand years ago.

The lack of earlier evidence of domestication in Egypt or elsewhere

*Bearing in mind the hybridization caveat from the last chapter.

supports the hypothesis that domestication occurred in Egypt within the last four millennia. But as we always say in science, absence of evidence is not evidence of absence. It's certainly possible that cats were domesticated somewhere else earlier before arriving in Egypt four thousand years ago. Or perhaps the first stages of domestication occurred elsewhere, and then when cats arrived in Egypt, they were transformed from human associates to the pets we know and love.

Unfortunately, two problems make it difficult to evaluate these ideas. First is the scant zooarchaeological record of cats or other evidence of human-cat associations prior to four thousand years ago.

Second, as we've already discussed, wildcats and domestic cats are nearly indistinguishable anatomically. Consequently, we can't look at the few skeletons that have been found of thousands-of-years-old buried cats and determine whether they were North African wildcats prowling the fields or already domesticated cats living the good life curled up by the fire. Furthermore, even if remains found in an archaeological site can be identified as those of a North African wildcat, there's still the issue of determining whether that feline was a wildcat that had been tamed or was from a population that had already begun to transform genetically on the road to domestication.

With those caveats in mind, let's examine what we do know.

Remember that ninety-five-hundred-year-old grave in Cyprus? The burial context and condition of the cat suggest an intimate human-cat relationship. Could this have been near the dawn of cat domestication?

The fossil record and older archaeological sites reveal an absence of cats in Cyprus before eleven thousand years ago (the discovery of a cat toe bone revealed the earliest evidence of cats on the island). Consequently, the buried puss must be the descendant of cats brought there by people, probably from nearby Turkey or Syria. The combination of cats being transported overseas and treated as a treasured family possession at death suggests—at least to some—that domestication had begun almost ten thousand years ago.

In reality, however, the evidence is not that strong. The ancient Cypriots also brought other definitely undomesticated species to Cyprus around the same time, such as foxes. Moreover, we've already discussed the difficulty in distinguishing between a tame wildcat and a domesticated cat. At face value, the ancient grimalkin could have been either.

Cat remains are found surprisingly infrequently in archaeological excavations from this period. The explanation for their paucity is unclear. Perhaps because they weren't eaten or utilized in any other way, they tended not to find their way into burials or refuse pits. Or maybe they were truly rare. We'll probably never know.

The next oldest notable record is a cat tooth from about eight thousand years ago, found in the settlement of Jericho. The first cat from Egypt was found at a site two thousand years younger. Like the Cyprus burial, it involved a cat buried at the foot of a human, in this case a craftsman buried with his tools and a gazelle (for sustenance in the afterlife?).

Much more exciting was the discovery of six cats—two adults (a male and a female) and four kittens—in a cemetery in Hierakonpolis, Egypt, dating to fifty-eight hundred years ago. Curiously, the female was only about six months older than the kittens. Domestic cats can reproduce at six months of age, so that in itself isn't extraordinary. However, wildcats in North Africa naturally reproduce only in the spring; hence, six-month-old wildcats would only occur in the fall, outside the breeding season. Domesticated cats, by contrast, breed year-round. Could this six-month offset be evidence that these were domestic cats? Tantalizing, but far from definitive. In these and a number of other Egyptian finds, the question always remains: was the cat tamed or had cats by this time been domesticated? Unfortunately, there's no way to know from looking at the bones.

And that brings us to the time of the pharaohs. The ancient Egyptians are famous for their love of cats. Felines first appeared in Egyptian art, hieroglyphics, and jewelry—even the use of "miit" (the Egyptian word for cat*) as a girl's nickname—about four thousand years ago. One tomb excavation from this period revealed seventeen cat skeletons next to small pots that may have contained milk. Whether they were tamed or domesticated was not clear, but some of these cats were clearly being cared for.

Starting in the reign of Thutmose III thirty-five hundred years ago, cats became very common elements in the decoration of tombs. They were often portrayed wearing collars, necklaces, and earrings; eating out of dishes; sitting lovingly under the woman of the house's chair or occasionally in the lap of her husband; even joining the family on hunting outings in the marshes. By this time, it was clear: the domestic cat had arrived.

The next fifteen hundred years were a glorious time for Egyptian cats (though with a dark side we'll get to shortly). About three thousand years ago, the ancient city of Bubastis in the Nile delta rose to prominence, and with it, the goddess Bastet. Portrayed for the previous two millennia as a woman with the head of a lioness, Bastet's cranium downsized from leonine to pussycat. In some sense, that wasn't such a radical change, because determining whether a feline head in ancient Egyptian paintings and statues is meant to be a lioness or a domestic cat can often be difficult. The transition from lioness to

---

*Sometimes spelled "miw," "miu," or a variety of other ways and reportedly translated as "he or she who mews."

cat may have occurred to emphasize Bastet's protective and nurturing persona as opposed to the fierceness of the contemporary lioness-headed goddess Sekhmet. For a time, the two were viewed as opposites—fury versus friendly. Not coincidentally, Bastet's qualities—playfulness, motherliness, fertility—are also those associated with domestic cats.

Cats connected with Bastet lived the good life. Some on the temple grounds were considered the incarnate manifestation of Bastet herself; the people worshiped not the cat himself, but the goddess embodied within him. Other cats did not achieve this designation, but were treated well simply because they were members of a sacred species. Egyptian religion had many deities with animal associations, but Bastet became the most revered of them all.

Beautiful bronze statues—notably the iconic upright cat sitting on its haunches—were made in industrial quantities. Family members shaved their eyebrows to mourn a cat's passing. Enormous cat cemeteries were created for the interment of pets and temple occupants. To kill a cat was a crime potentially punishable by death.

Eventually the Egyptian dynasties crumbled and the Romans invaded. The old religions withered, Bastet's star faded, and cats once again were just cats. By that time, however, the feline conquest of the world was well underway. We'll get to the great feline diaspora shortly, but first there's one more ancient cat archaeological find I need to discuss, from a time before the Egyptians had domesticated cats and from a most unexpected place.

CATS ARE FIRST RECORDED in China about twenty-two hundred years ago in court records from the early Han dynasty. Imagine the surprise, then, when archaeologists excavating a fifty-three-hundred-year-old village in central China found eight cat bones from at least two average-size cats.*

Just as in other old archaeological sites elsewhere, the challenge was to figure out what the cats were doing there. Were they wild cats whose bones just happened to end up in the village—possibly even hunted for food or fur—or did they have some sort of commensal relationship with the villagers?

Researchers took a novel approach to addressing that question: they tried to determine what the cats were eating. To figure that out, the scientists took advantage of the fact that you actually are what you eat: foods differ in their chemical makeup, and organisms incorporate molecules from the food they eat into their own bodies. Thus, by analyzing an animal's chemical makeup, researchers can draw inferences about what was on their menu.

Many agricultural crops have a different type of carbon (technically, a different carbon isotope) than other types of vegetation.† As a result, animals that eat agricultural crops have higher concentrations of one carbon isotope than animals feeding on the vegetation growing wild outside the agricultural fields. The bones of humans, pigs, dogs, and a rodent from the village all had the isotopic signature of millet, a common crop grown in that region at that time. By contrast, the isotopes of several deer bones indicated a diet of non-agricultural plants.

The cat bones showed an isotopic signature more similar to that of the village dwellers, indicating a substantial contribution of millet to

---

*The minimum number of individuals is established by duplicates of the same bone, in this case the presence of two left tibia (the bone in the lower leg). The bones were found in three garbage pits, suggesting that they may have come from three individuals.

†As a result of the possession of C3 versus C4 photosynthetic pathways, for biochemistry aficionados.

their diet, too. One possible explanation was that the cats were eating rodents that feasted on millet, but isotope levels of another chemical element, nitrogen, indicated that their diet was composed of both plant and animal matter; one cat in particular had a diet dominated by plants with little evidence of animal protein consumption.

Cats are hypercarnivores in nature. Unlike dogs, cats in the wild eat nothing but animal flesh. The fact that these Chinese cats were getting some or all of their sustenance from plants could mean only one thing: they were either being fed by people or eating village scraps (just as Darwin noted for a French kitten).

There was a second clue as well. The teeth on a jawbone were very worn down, indicating that the cat was very old. Although animals in the wild sometimes survive until they are decrepit, a more likely explanation in this case is that people were caring for the cat.

These findings suggested that at least the earliest stages in the domestication process were occurring in China much earlier than anyone had expected, before cats had been fully domesticated in Egypt.

But what kind of cats were these and where did they come from? The authors speculated that either the cats were North African wildcats transported from the west to China, for which there was no evidence, or that the animal was an Asian wildcat following the same domestication pathway as its North African cousin, a possibility for which there was also no evidence.

Then the plot thickened. A French zooarchaeologist, the same one who had studied the ancient burial on Cyprus, became intrigued and decided to take a closer look. In collaboration with Chinese museum curators, he examined the specimens, as well as skeletal remains from three other cats discovered at two other equally ancient sites elsewhere in China.

One of these other archaeological specimens was a complete cat skeleton laid on its side. As with the Cypriot cat, the careful entombment suggested that the cat was not a wild cat living around the settlement, or even a meal, but a family member receiving full burial honors—yet more evidence that cats were somewhere on the domestication pathway in ancient China.

Still, what kind of cats were these? The researchers focused on the jawbones of four of the cats. They took careful measurements of the lengths, widths, and angles of every aspect of the jaws, comparing the ancient bones to museum specimens of domestic cats, wildcats, and the Chinese leopard cat (another local species, Pallas's cat, was ruled out because that species has a characteristic configuration of the bones around the ear not seen in the archaeological specimens).

The shape of the back end of the jaw—the part that juts upward compared to the horizontal shape of the rest of the jaw, just as in humans—turned out to be the key. In leopard cats, this part of the jaw is very upright and somewhat larger, whereas in wildcats and domestic cats, the projecting bone leans backward and is a bit smaller.

Comparison to the archaeological cats was clear-cut. All four had bony protrusions standing straight up. They were definitely leopard cats, not wildcats or domestic cats!

This result was a shock. The established wisdom was that cats had only been domesticated once, somewhere in the Middle East, from North African wildcats. This finding stood that result on its head.

The leopard cat (remember, a small, spotted cat not closely related to the leopard) is a candidate for domestication in some ways, but not in others. Widely distributed in Asia, these cats are much more tolerant of human activities than most small feline species; they are often abundant today in disturbed habitats such as agricultural fields and oil palm plantations, where they may play an important role in controlling rat populations. Although they mostly stay away from humans, there are reports of them occasionally entering villages, eating garbage, and preying on chickens. It's not much of a stretch to imagine leopard cats moving from such habits to becoming regular scroungers around Chinese villages, though I am unaware of any such observations from modern-day settings.

An Asian leopard cat.

On the other hand, there's reason to be surprised at the prospect of leopard cat domestication. Remember the zookeeper survey about the affiliative behavior of different feline species mentioned in chapter three? The African wildcat was among the friendliest species. No wonder it was domesticated. It was already friendly to start with.

And which species was at the very bottom of the list, the unfriendliest of the unfriendly? None other than the leopard cat.

Given this temperament, I question how domesticated or even tamed these archaeological leopard cats really were, even if one was buried carefully. People aren't going to warm up to an animal that doesn't reciprocate in any way. The cat domestication pathway involves a coevolutionary walk, the back-and-forth of cats being at least a little friendly, people providing some food and hospitality, cats being selected to be friendlier, people reciprocating with more food and shelter, and eventually you have a lap cat and devoted human fans. This coevolutionary give-and-take seems much less likely to happen if the cats don't give a little love in the first place.

Of course, we'll never know whether these Chinese cats were tamed or domesticated. Regardless, what happened to them? If domestic cats had interbred with them when they reached China, then we would expect to find evidence of leopard cat DNA in modern domestic cats, just as Neanderthal DNA now exists in modern human populations. But no such evidence has turned up. Leopard cats did not contribute to the gene pool of today's domestic cats.

So leopard cats may have been commensal with humans, hanging around villages, eating rodents, maybe getting the occasional handout. But the spotted cats likely didn't follow their North African relatives down the domestication road. And when the domestic cat was brought to China two thousand or so years ago, they took over the village cat niche, perhaps because they were more affectionate or better mousers, or for some other reason.

PALEONTOLOGY AND its sister discipline, zooarchaeology, rely on real, concrete evidence of what organisms were like in the past. When you find the massive femur of a *Brontosaurus,* no one can (reasonably) deny that enormous beasts walked the earth in the Jurassic period. But ancient material is often rare and degraded. Many fossil species are known from only one or a few specimens, sometimes from a single bone or even a footprint! We've seen how scarce cats are in archaeological deposits. There aren't a lot of data to work with.

The alternative approach is to reconstruct the past from the present. An evolutionary biologist can build an evolutionary tree of modern genealogical relationships and then use the phylogeny to infer evolutionary history. For example, both giant and red pandas subsist on bamboo and possess thumb-like structures to manipulate the tall, tubular stalks. Scientists used to think that the two species were closely related, their similarities the result of descent from a common, bamboo-munching ancestor. However, the phylogeny indicates a different story: giant pandas are a type of bear, whereas red pandas are more closely related to raccoons. As a result, the two pandas must have evolved their similarities independently; their false thumbs are the result of convergent adaptation to the same environmental situation.

In the old days, scientists used anatomical data such as the shape of the jaw protrusion in leopard cats to infer phylogenetic relations. However, collecting such data is painstaking, requiring poring over specimens to identify sometimes minute differences in anatomical structures. It takes a long time, and you end up with data for relatively few different characteristics.

Today, most biologists use DNA data to construct phylogenies. The advantage, as I've already mentioned, is that large quantities of such data can be collected easily and quickly.

In an ideal world, the two approaches—examining ancient bones and DNA—would be combined. A great opportunity to do that was provided by the Chinese cat bones. Why didn't the archaeologists just do a DNA test on the Chinese cat bones to see if they were from a leopard cat? The answer is that it's easier said than done.

Eight

Curse of the Mummy

When I lecture on fossil lizards entombed in amber, I ask my audience if anyone has *not* seen *Jurassic Park*. Very few people raise their hand. The dinosaur sci-fi thriller is nearly a cultural universal. Few realize, however, that not only is *JP* great fun to watch, but in scientific terms, it was well ahead of its time. As you'll likely recall, the movie's premise was that dinosaur DNA could be recovered from blood-sucking insects encased in amber for more than sixty million years, an idea that seemed fantastical. Thirty-plus years later, however, what was then science fiction is now, at least in part, science fact.

Scientists have succeeded in extracting intact DNA from many ancient specimens. Dinosaur-vintage fossils appear to be too old—DNA eventually disintegrates over many millions of years—but younger specimens often yield intact DNA: million-year-old woolly mammoths, half-million-year-old horses, even four-hundred-thousand-year-old bones from our Neanderthal relatives. Temperature and humidity seem to be the key: the colder and drier the better, which explains why few tropical specimens have proven useful and many of the successes have come from the frozen north.

The study of such ancient DNA has revolutionized archaeology. Sparked by the analysis of DNA from Egyptian mummies in the mid-1980s, the genetic study of thousands-of-years-old human remains has become a vibrant area of research.* By examining the DNA contained within bones, teeth, skin, and even chewing gum, new insights have been gained about how modern humans spread over the world, how humans and Neanderthals interacted, even what ancient people ate and what illnesses befell them.

Not surprisingly, many of the sharpest young minds have been drawn to this intersection of biology and archaeology. One of them is Claudio Ottoni, the son of a biologist who became entranced in college with how modern molecular genetics could help decipher the human past. For his doctoral work at the University of Rome Tor Vergata, he focused on ancient nomadic pastoralists in the central Sahara Desert. A rich archaeological record exists from this region, including beautiful rock paintings and evidence of early animal enclosures. It is unknown, however, whether the people who live there now, the Tuaregs, are descendants of these early desert dwellers. Ottoni hoped to find out by

*Svante Pääbo, who conducted the mummy study, won the Nobel Prize in Physiology or Medicine in 2022 for pioneering this field.

comparing the DNA of modern Tuaregs to that of skeletons unearthed from Libyan archaeological sites.

The ancient DNA portion of the project was a failure. No usable DNA was recovered, probably, Ottoni conjectured, as a result of the high temperatures the skeletons endured during their six-to-ten-thousand-year interment in the sands of the Sahara. Yet, in all other respects, the thesis was a grand success. The analysis of modern DNA provided great insight on the origin of the Tuareg in very much the same way that Driscoll's study illuminated the origin of domestic cats. Moreover, despite the failure to obtain ancient DNA, Ottoni proved in a set of technical papers that he had mastered the techniques involved. It wasn't his fault that the DNA had baked away before he had a chance to examine it!

Ottoni then landed a prestigious postdoctoral position at the Laboratory of Forensic Genetics and Molecular Archaeology in Belgium, where he worked with renowned zooarchaeologist Wim Van Neer. Although best known for his work on fish, Van Neer had published several papers on cats, including the study reporting six cat skeletons from Hierakonpolis.

Ottoni's job was to bring a genetic perspective to the work of Van Neer and the other scientists at the center. The actual plan was vague. At the outset, Ottoni was keen to continue working on ancient human DNA. But the more he learned about zooarchaeology, the more benefit he saw in switching species. The study of ancient human DNA is a crowded and contentious area. Why not move to a topic with a bit more elbow room and less drama? Ottoni split the difference by initially working on both ancient humans and ancient pigs.

The porcine project was a big success. By sequencing DNA from pig skeletons from forty-eight archaeological sites in western Asia, the oldest dating to twelve thousand years ago, Ottoni showed that the history of pig domestication was much more complicated than previously realized. The paper was well received in the scientific community and helped further establish Ottoni's reputation.

Repeatedly during Ottoni's social visits to Van Neer and his house full of cats (living, not archaeological), the senior scientist would try to get Ottoni interested in ancient cat DNA. "What about cats? I have the samples," Ottoni recalls the senior scientist inveigling. Ottoni kept putting him off for one reason or another.

In reality, Ottoni was scarred by his experience with the Tuareg archaeological samples. Much of Van Neer's cat material also came from desert areas in northern Africa. Ottoni was not eager to go down that path again, fearing the result would be the same: no usable DNA. But finally, after the pig paper came out, Ottoni relented. He'd give the cat samples a shot.

One of the first samples Ottoni sequenced was a cat from the Red Sea port town of Berenike, dating to about 150 AD, during the Roman rule of Egypt. The results showed that the DNA from the Berenike sample appeared to be from an Asian wildcat. Finding its DNA in Egypt made no sense to Ottoni; the nearest Asian wildcats today are in southwestern Asia, on the other side of the Arabian Peninsula. He wrote the result off as a mistake of some sort and didn't give it another thought.

Then Van Neer pointed out that two thousand years ago, Berenike was a major locus in trade from South Asia to the Mediterranean basin by way of the Red Sea. Suddenly, the presence of a cat with Asian DNA in an Egyptian Red Sea port made a lot more sense. Moreover, the finding suggested a very exciting possibility: perhaps two thousand years ago, people in India or elsewhere in Asia were trying to domesticate the Asian wildcat.

Ottoni was energized. Not only could these samples yield usable DNA, but they might provide unexpected results. He turned his attention to the rest of Van Neer's samples. The zooarchaeologist's wide network of collaborators had uncovered cat remains at sites spanning most of Europe, western Asia, and northern Africa, dating from the Stone and Bronze ages, the Roman and Byzantine empires, even a medieval Viking trading port in northern Germany. Still, there was one glaring hole in Van Neer's web.

When ancient Egypt is mentioned, two topics usually come up: the pyramids and mummies. In fact, the two are related, because the pyramids are essentially enormous tombs containing the mummified bodies of dead rulers.

Egyptian mummies—skillfully eviscerated, dehydrated, chemically treated, and wrapped in linens—are a scientific marvel. The decorations applied to their wrappings and surrounding cartonnage rightfully earn them a place in art museums the world over. But most people know Egyptian mummies from movies featuring ancient curses and resurrected rulers with supernatural, malevolent powers.

Even though the 1999 remake of *The Mummy* spawned two sequels, one prequel, four sequels to the prequel, and an animated television series, Hollywood has never taken advantage of what would seem to me to be an obvious plotline. The Egyptians not only made mummies of humans but also of a wide variety of animals, including lions, crocodiles, baboons, and snakes. Surely this zoo-mummological profusion provides great opportunity to expand the typical mummy movie fare beyond the customary ancient curses and revivified rulers.

When I delved into this topic, I expected to learn that cats held the top spot in the Egyptian embalmed menagerie. To my surprise, that's not the case. Though they receive much less attention, dogs were the most mummified species—one catacomb (dogacomb?) contained nearly eight million mummified canines. But in second place?* Cats, by the millions. Surely *The Mummy Meowed* would be box office gold.†

Perversely, the death and subsequent mummification of so many felines was the result of the revered status of cats in ancient Egypt. During Bastet's heyday, hundreds of thousands of Egyptians would make the pilgrimage to the great annual festival in her honor in Bubastis. And once they got there, in between Mardi Gras–like debauchery, revelers would visit the temple complex.

As a way of paying homage to the goddess or requesting that a prayer be answered, pilgrims purchased mummies, which they left as votive offerings in the temple complex. Different Egyptian gods had different totemic animals; cats were the totem of Bastet, so worshippers presented mummified cats. Presumably, the larger the ask, the more elaborate and well preserved the proffered mummy.

Once a year, priests took the mummies down into the (appropriately named) catacombs for storage. Millennia later, when these subterranean galleries were excavated, the abundance of mummies was so great that tons of them were shipped to England to be ground up and used as fertilizer.

In ancient times, the sale of these mummies on a vast scale may have been a moneymaking venture that supported the temple complex. So lucrative, in fact, that sometimes the cat mummies did not actually contain cats—at one site, one third of the mummies contained only mud, clay pebbles, or in one case, a mummified fish—presumably because the supply of dead cats ran short.

X-ray analysis of the mummies (trigger warning!) reveals that most

*Or maybe third. Ibises—wading birds with curved beaks—were also embalmed in great numbers, but no accurate estimates exist of how many.
†Subsequent to writing this, I learned that cat mummies were, in fact, featured in a 2019 online streaming addition to the *Scooby-Doo* series.

were either large kittens or young cats, and that they were killed by breaking their necks or by strangulation. Presumably, vast catteries existed on the temple grounds or nearby where the cats were bred and killed to produce the mummies. Archaeologists have not yet discovered evidence of them, but similar complexes for breeding birds and even crocodiles have been found.

For a long time, these mummies were treated more as an annoyance than as objects worthy of archaeological study, and as a result, there are fewer specimens in museums than one might expect. For example, of a nineteen-ton shipment of 180,000 cat mummies sent to Britain in the late nineteenth century, only a handful ended up in a museum.

Nonetheless, enough found their way into collections for an aspiring zooarchaeologist to see the opportunity to obtain ancient Egyptian cat DNA. And so Ottoni requested permission from the Natural History Museum in London to take small pieces of bone, hair, and skin from the mummies to extract any DNA they might contain.

As a former curator myself, I can attest that museums look very skeptically on requests for what we call "destructive sampling." Our specimens are precious and often irreplaceable; any research that involves damaging them in any way needs to have very strong justification.

Luck, however, was on Ottoni's side. A previous study had used an X-ray machine to visualize what was inside the cat mummies' wrappings. That's how the scientists had learned the cause of death (and also how they'd discovered that occasionally something other than a complete cat was wrapped in what externally looked like a cat). However, the X-ray machine the museum owned was rather small.

Cat mummies come in two forms. In some, the cat was mummified in a normal cat shape—the form of the mummy was clearly recognizable as a cat. In other cases, though, the mummy was cylindrical in shape because the cats were preserved with their legs pressed against the belly (forelegs stretched back, hindlegs folded in sitting position, tail tucked between the hindlegs against the belly).

The British museum's mummies were of the latter type and were too long to fit in the X-ray chamber. In a decision that shockingly contradicts what I just told you about the sanctity of museum specimens, the previous researchers had been allowed to decapitate the mummies so they could fit inside the machine. This may have been bad for the specimens' integrity, but it helped Ottoni's case. The damage had already been done; instead of needing to unwrap and dismember a mummy, he was only asking permission to reach inside and pluck out a piece of exposed skin or bone, a request that was eventually granted.

Ottoni's voice softens as he recalls his time there, one of the highlights of his career. Deep in the bowels of the enormous museum, the basement room itself seemed like a bit of ancient history: packed with archaeological specimens and very silent; no phone or other modern contraption. And the smell—old, stale, a little funky. Wondrous antiquities—monkey mummies!—stashed in museum cabinets, tucked into every nook and cranny.

But the cat mummies are what stand out in his memory. Some clearly had been prepared with little care, the preparation rushed. But others were exquisitely swathed, strips of linen elegantly interwoven, the wrappings around the head carefully molded into the proper feline countenance, pointed ears attached on top. Eyes were painted onto the linen, sometimes in red, along with the nose and other facial markings. Viewing these objects, Ottoni was transported back in time three thousand years to a different culture, a different way of life.

As I've already mentioned, one problem with accessing ancient DNA is that it is often highly degraded, and thus the quantity of usable DNA is often far less than what you'd get from fresh material. As a result, one risk is that your samples will be contaminated by modern

DNA. Scientists have come to realize that the world is full of DNA shed by organisms (including you!) floating around in the air, on surfaces, in water. If you have a lot of DNA from your specimen, then it doesn't matter if a strand of someone's or something else's DNA lands on the specimen's surface; its presence will be swamped out by the millions of copies of the specimen's DNA. But if you can only get a tiny bit of DNA from the specimen, or even none at all, then the contaminant DNA can constitute a major portion of what you get. So when you run the analyses, the DNA sequence you get back is likely to be from the contaminant DNA, not from the specimen you're trying to study.

Of course, this is only a problem if the DNA that lands in your sample is from the same species or a closely related one. If you're studying an ancient cat and the DNA results come back and say the specimen is a toad, you know something's gone wrong. So if you're studying an obscure species—a duck-billed platypus, for example—it's unlikely that there will be modern platypus DNA floating around (unless you're in Australia). But if you're trying to get cat DNA (or, even worse, human DNA), it's a different story—you've got to be very careful.

Ancient DNA researchers have adopted a series of sterile procedures to prevent contamination. When obtaining a sample from a cat mummy, Ottoni and Van Neer wore rubber gloves and face masks, putting on new ones for each specimen to prevent cross contamination. All their tools and the tabletop were cleaned with bleach and acid solutions after each specimen for the same reason. Bone and skin samples were carefully wrapped in aluminum, then placed in a plastic bag for transport.

Once a sample arrived back at the DNA lab, it was moved to a sterile room. Workers who entered were covered head to toe in protective gear. The sample was cleaned with bleach. Then the outer surface was removed with a razor blade or drill, uncovering parts of the tooth or bone that had not potentially been exposed to DNA contaminants (hair and skin samples were more difficult to sterilize). Material was then carefully removed, ground to powder, and subjected to chemical treatment to extract the DNA. This entire procedure was repeated multiple times, and the DNA from the different trials was compared; the same foreign DNA would be unlikely to contaminate multiple extractions.

And Ottoni had one more safety precaution. Despite his girlfriend's entreaties, he had not gotten a cat. Too easy to inadvertently carry modern-day cat DNA into the lab (as Ottoni pointed out to me, cat DNA is actually used as a forensic tool in criminal investigations because it is so easily transported by cat owners; one murderer was convicted largely thanks to hairs from his cat, Snowball, found on a jacket spattered with the victim's blood).

OTTONI'S GOAL WAS TO USE ancient cat DNA to document the history of cat domestication and geographic spread. His samples varied not only geographically, but also chronologically. He hoped that by tracking the spread of particular DNA variants (alleles) through space and time, he could map the spread of domesticated cats, revealing once and for all the route by which cats conquered the world.

There were, essentially, three hypotheses. The traditional view was that cats became truly domesticated—more than just commensals—in Egypt by thirty-five hundred years ago, and then spread from there in all directions to the rest of the world. A vocal minority, however, argued that Turkey or thereabouts was where domestication occurred, spreading from there to Egypt and elsewhere. Evidence for that hypothesis came from the archaeological cat in Cyprus, not far from Turkey, and the fact that cats are beloved in Turkey today. Finally, a third possibility was that domestication occurred in both places.

The way to test the out-of-Egypt hypothesis is clear: look for an allele that initially is found only in cats in ancient Egypt but occurs at younger, more recent sites elsewhere in the world. Such a pattern would indicate that Egypt had been the ancestral source of cats.

It took Ottoni four years to work through all 352 samples. The results, however, were more tantalizing than conclusive. Just as with his doctoral work with the Tuareg, Ottoni was vexed by samples that failed to yield usable DNA. His overall success rate was a quite impressive fifty-nine percent, but the British Museum mummies were a big exception: only six of seventy-four bore DNA fruit. Very likely, the embalming process, involving dehydration and treatment with various salts and chemicals, played a role in degrading the DNA; the hot Egyptian climate didn't help either.

The lack of data from the oldest Egyptian specimens turned out to be critical. Ottoni did find an allele that appeared in Egyptian mummies twenty-five hundred years ago and then, a thousand years later, became common throughout Europe. Unfortunately, the allele also appeared in Jordan, Turkey, and Bulgaria at more or less the same time as in Egypt. In theory, data from older specimens could pinpoint where the allele first appeared, and thus illuminate the path by which it spread elsewhere, but that will require more work. Getting data from older mummies would be particularly helpful.

Unexpectedly, the most exciting finding came from ancient Europe. The European wildcat had been thought to range throughout all of Europe and into Turkey, but Ottoni's data said otherwise. None of the seventy-two Turkish samples contained European wildcat DNA. Three had Asian wildcat heritage (more evidence of an attempt to domesticate that subspecies?) and the rest were North African wildcats. True to its name, the European wildcat's range was, indeed, limited to Europe.

This result may seem a little underwhelming. Scientists got the historical distribution of wildcat subspecies a little bit wrong—so what? It's always good to set the record straight, but does this new knowledge have any significance for understanding the origins of the domestic cat?

Indeed it does. Samples from younger archaeological sites revealed that an allele previously found in North African wildcats in Turkey arrived in southeastern Europe by six thousand years ago. Notably, people had taken the same route migrating to the region from Turkey, bringing agriculture with them. Subsequently, a Polish team found that cats bearing the Turkish allele were present in central Europe about the same time as agriculturalists appeared there as well, nearly five thousand years ago.

The implication was that two thousand years before Egyptian finishing school, Turkish cats of the North African wildcat lineage accompanied human immigrants from Turkey to Europe. If these cats made the trip as pets rather than commensals, then the out-of-Egypt hypothesis would be disproven, at least as the sole explanation for domestic cat origins. The other possibility, of course, was that the migrating cats were still commensal hangers-on, tracking the advance of agriculture and the bounty of rodents that comes with it.

The key question, then, concerns the relationship of the immigrant cats to the people with whom they associated. We've already seen how to answer that question: isotopes! The researchers who discovered ancient North African wildcats alleles in Poland employed the same isotope chemical method previously used in the Chinese leopard cat study to determine what the cats were eating.

The analysis, however, was slightly different. Animals—including humans—that eat crops grown in fields fertilized by manure have higher concentrations of a particular nitrogen isotope than animals feeding in unfertilized areas. In turn, predators that eat prey that fed on these crops will also have higher levels of that isotope. Of course, this approach can't answer all the questions—a cat could be eating prey in the agricultural fields whether it was a pet or a commensal—but it can give a sense of how closely associated the cats were with people.

The Polish scientists made two comparisons. First, they compared the isotopic nitrogen levels in the bones of the immigrant cats they had excavated to the comparable level in European wildcats from an earlier, preagricultural period in Poland. Their reasoning was that if the newcomers were eating prey from farmed areas, they'd have heightened levels of the nitrogen isotope compared to the non-agricultural European wildcats. Second, they compared isotope levels in their immigrant cats to those of people and their dogs living in the same area at the same time. If the cats were living as domesticated animals, they reasoned, their levels should be comparable to the humans and the dogs.

The results placed the Polish cats squarely in the middle. Their isotopic levels were higher than the preagricultural European wildcats, but lower than the dogs and the humans. The interpretation was that the cats were hanging around people, feasting to some extent on the rats and other critters feeding in the agricultural fields, but they were not living full-time as domestic family members, as the dogs were.

Of course, cats are not dogs. They don't eat the same foods. Perhaps the difference between the cat and dog isotope levels had more to do with what they ate, rather than how domestic they were. The Polish scientists had an answer for that, too. They also measured the isotopic levels in dogs, cats, and humans from a more recent Polish site, an agricultural area dated to two thousand years ago, when the cats were much more likely to be domesticated. In those samples, the isotopic levels of the cats were about the same as the dogs and the people, and substantially higher than the older cat samples. Isotopic nitrogen levels, in other words, seem to be a good marker of living in domestic harmony with humans.

The study leads to a very important conclusion: five thousand years ago, Polish cats, descended from felines that had migrated from Turkey, were human associates, living around farming communities but in an undomesticated, commensal lifestyle. This scenario jibes with the archaeological context of the cat skeletons, which were found in nearby caves, rather than in direct association with humans.

These ancient Polish cats provide tangible proof for a critical step in the standard model of domestication: cats hanging around humans benefiting from our presence and probably providing a rodent-control benefit in return.

Yet, domestication didn't occur. The ancestors of the Polish cats had been living in coexistence with humans in the Fertile Crescent since the dawn of agriculture, but these data suggest they still were only commensals, not domesticates. Think about animals common around human settings today—raccoons, opossums, pigeons. These animals are not domesticated, but as human settlements move, the animals move with them. North African wildcats may have done the same thing. Something more had to happen for wildcats to cross the threshold to domestication.

We still don't know what that something was, but conventional wisdom is that it occurred in Egypt, perhaps about four thousand years ago. And once cats were domesticated, the story goes, they became much more desirable and quickly spread to the rest of the world.

Admittedly, this out-of-Egypt scenario is based mostly on the paucity of archaeological evidence of domesticated cats elsewhere. As I said earlier, absence of evidence isn't evidence of absence. Perhaps cats were domesticated in what is modern-day Turkey, Syria, or elsewhere thousands of years before they were painted on tombs in Egypt, and we simply haven't found proof of it yet. But the evidence from the Polish site is one strike against this hypothesis; if cats had been domesticated in Turkey, the ancient Polish cats should have shown signs of it.

Enough for now about ancient DNA. We'll return to what studies on modern cats can tell us about the feline diaspora in chapter fifteen.

THE SPREAD OF CATS from Egypt to the rest of the world is fairly well documented from archaeological sites and historical records. The Egyptians tried to prevent the export of cats, allegedly going so far as to have their armies, when on military expeditions, retrieve all captive cats in the invaded country and return them to Egypt. But it was to no avail. Sailors saw their value for rodent control and smuggled them on board. Phoenician traders plying the waters of the Mediterranean—supposedly referred to as "cat thieves" by the Egyptians*—conveyed them away, as did ships from Egypt and elsewhere.

Why cats were desired elsewhere is not clear. The Greeks and Romans already employed ferrets and weasels as ratters, and thus had no particular need for cats to fill that role. At least in Rome, they weren't initially popular as pets either. It's possible that their initial spread was solely a result of maritime activity—sailors valuing cats for their pest-control activities—which served to spread them around the Mediterranean, and then they were on their own once they abandoned ship in foreign lands.

In Greece, there is evidence of cats from frescoes, jugs, cups, stone seals, and daggers going back as far as 1700 BC. Domestic cats were certainly established by the fifth century BC in Greece, if not before,

*Though I've been unable to track down the source for this claim.

and from there they quickly moved throughout much of southern Europe. The Romans then played a major role in cat dispersal to much of the continent, an interpretation based on the fact that in many areas the earliest evidence of cats is found in Roman settlements.

Cats also spread east throughout Asia by both land and sea. I've already talked about the overseas trade route through the Red Sea to South Asia that probably brought cats to India. In addition, cats traveled overland, hopping onto the Silk Road to Iran and from there eastward all the way to China. A sculpted relief in Pakistan shows a domestic cat quarreling with dogs in the third century BC; within a few hundred years, evidence of cats crops up in India and China. By 600 AD, they were in Japan. All were descendants of North African wildcats.

THERE ARE LOTS OF COOL noteworthy findings in Claudio Ottoni's work, but when the study was first presented at a conference in Oxford, one particular discovery captured the world's imagination. No, not cat mummies, though that would have been my guess. Nor was it cats invading Europe from Turkey, nor cats involved in Red Sea trading with Asia.

It was Viking cats that went viral.

Renowned for their seafaring prowess, the Norsemen brought cats along on their voyages for rodent control and—who knows?—maybe companionship. Some scientists had speculated that Vikings were the ones who brought cats to northern parts of Europe, as well as to Iceland and other islands.

Ottoni's study provided support for this idea. One of the archaeological sites in his study was a seventh-century AD Viking trading port on the Baltic Sea, in what is now Germany. Finding that the cats in this village had the same allele as Egyptian cats, Ottoni suggested the Vikings had, indeed, picked up the cats in the Mediterranean and transported them home by longship.

The world's media ate it up. "Cats Sailed with Vikings on Their Quest to Conquer the World," said *Business Insider* (that's right, a business journal covering cat science). "Did the Vikings Help Distribute Cats around the World?" asked *The Christian Science Monitor.* Articles appeared in newspapers, magazines, and websites, large and small, the world around. Memes of cats wearing Viking helmets, some on the prow of a ship Kate-and-Leo-style, popped up all over the internet. "I didn't even know there were Viking cats," an embarrassed geneticist was quoted as saying in many different articles.

The geneticist can be forgiven for his educational lapse. The Vikings had cats, but so did many other people throughout history—it's not clear that Viking cats were a real thing. There are many interesting and fanciful ideas about Vikings bandying about the internet—and even the scientific literature—and it pays to be skeptical. Much of the way human Vikings are popularly portrayed is not correct. Helmets with a horn on each side? They never existed, much less were they worn by cats.* With that in mind, let's take a closer look at what we know about cats and Vikings.

Adherents of the view that the Vikings venerated cats often cite as evidence Freyja, the mythological Norse goddess of love, beauty, and fertility, who supposedly traveled on a chariot drawn by two cats, Bygul and Trjegul. However, recent scholarship suggests that Freyja was actually said to have been riding on a boar, and the cat story was revisionist

*What? You didn't know that? No horned Viking helmets have ever been discovered. They were the creation of Swedish artists in the nineteenth century, made famous by the production of Wagner's Ring cycle operas in the 1870s and a staple of Viking imagery ever since. I didn't know that either until I started researching this topic.

mythology invented later by the Christians. Even worse, the names of Freyja's cats were not—despite what is widely reported on the internet and elsewhere—Bygul and Trjegul. Those names were created in a 1984 fantasy fiction novel, *Brisingamen*, by Diana Paxson. What the cats' true Norse names were, if they had any, is lost in the mists of time.

Freyja aside, what's clear is that like the Egyptians, the Vikings' relationship with felines was complex. Cats were treasured as pets and buried with their owners, but also sacrificed in religious rituals and slaughtered in great numbers for their hides, which were used for lining and trimming clothes. We know the latter from many archaeological sites with cat bones bearing cut marks produced in the skinning process, including one in Denmark from 1070 AD that contained a pit filled with remains from at least seventy cats.

Incidentally, the oft-heard claim in cat documentaries that the Vikings particularly loved orange cats and are responsible for spreading them around Europe and elsewhere has no more factual basis than the horned Viking helmets. As far as I can tell, the entire evidence is that some islands north of Great Britain have an unusually large percentage of orange cats, which one research paper suggested might have resulted from Vikings preferring orange cats and taking them along when they established colonies there almost a millennium ago.

As THE STORY of the Viking cats makes clear, a lot of the details remain to be uncovered about how cats dispersed around the world. Ottoni is undeterred, however, and has just received a major grant from the European Union that will allow him to use more sophisticated genome sequencing methods to get much better genetic data. Combined with additional specimens that Van Neer has been collecting, the project should be able to provide a substantially more detailed record of cat origin and dispersal.

At the same time, Ottoni plans to take a second approach to understanding the history of cat domestication. Remember those thirteen genes that had previously been identified by genome analysis as being under selection during cat domestication? By examining entire genomes of archaeological cats, Ottoni hopes to pinpoint when and where these genetic changes first appeared. Such data should go a long way to settle the question of when *Felis silvestris lybica* became *Felis catus*. The project kicked off in early 2021. I expect you'll be reading about it in *The New York Times* shortly after this book is published.

Regardless, what is clear is that by fifteen hundred years ago, domestic cats were found most everywhere in Europe, Asia, and the north of Africa. The cat had conquered the world, yet its evolutionary journey to the spectacular diversity we know today had just begun.

Nine

Calico Tigers and Piebald Pumas

Painted cats in Egyptian tombs are usually doing one of three things: sitting with people, usually under a woman's chair; going along on a hunting expedition in a marsh; or fighting with a snake, often using a knife to good advantage. Some of these feline representations are more realistic than others, but there is one constant: in all of them, the appearance of the cat is the same: tawny yellow with vertical stripes or lines of spots, a ringed tail, and stripes on the limbs. The cats are also slender to moderate in build and often have long legs, a body shape paralleled in many sculptures from that time.

Ancient Egyptian cats, in other words, looked a lot like North African wildcats, though perhaps a bit lighter in color. That's hardly a surprise given their ancestry. What is interesting, though, is that there is no evidence of variation, no hint of the great variety of colors and patterns that we know in today's cats.

My personal history with cats exemplifies the vast potpourri of modern cats. Henry was the third cat I lived with as an adult. Stocky, gray with faint dark stripes down his side, black rings on his legs and tail, and the typical M mark on his forehead, he was one of those striped mackerel tabbies* that look more or less like a European wildcat.

None of my other eight cats past and present looked remotely like a wildcat, European or otherwise: two beautiful off-white Siamese rescue cats, Tammy and Mauritia, with black feet, ears, face, and tail; Nelson, with similar dark markings, set against a lustrous brown coat; Jane and her doppelgänger, Curaca, both medium gray with no markings; Leo, an orange Abyssinian with dark tickings at the end of each hair, giving a flecked appearance;† Archie, light orange with tabby markings on his face and legs, but nowhere else; and Winston, with his blotchy gray-and-white patchwork.

A mackerel tabby.

My feline novenary reflects but a fraction of domestic cat diversity: calicos, torties, black cats, and smokes; longhairs and shorthairs; plumy tails and no tails; snowy white on the body with other colors on

*In the cat world, "tabby" has a specific meaning, referring to a cat with the M on the forehead. Patterns on the flank vary from full stripes to stripes broken into lines of dashes to spots to nothing at all.

†Many mammals have similar "ticked" hair; the technical name is "agouti," after a tropical Central and South American rodent with hair of this sort.

the head and tail. Not to mention the many new looks developed by breeders in the last few decades. Remarkably, much of this medley of cat variety can be found in any free-living domestic cat population. Yet, none of it is seen in African wildcats today or in the cats of ancient Egypt. When and why did this cat kaleidoscope emerge?

REPRESENTATIONS OF CATS in ancient Greece and Rome are fewer in number than in Egyptian artwork, but they tell the same story—early domestic cats looked a lot like African wildcats. Taken at face value, this record indicates that today's diversity of housecat appearance did not originate until well after pusses departed their ancestral home and spread through Asia and Europe.

Non-tabbies start showing up in the historical record two thousand years ago. The first surviving pictorial representation of a non-wildcat-looking moggie is a mosaic tile from southern France illustrated with a black cat and thought to be from early Roman Empire times. Several centuries later, the Greek physician Aetius wrote about both black and white cats. Skipping ahead a few years, there definitely were black cats in twelfth-century Europe because we know there were superstitions about them; hysteria was then accelerated thanks to a papal decree in the thirteenth century that led to centuries of black cat massacre. White cats appear in some European paintings and frescoes in medieval times. A thirteenth-century encyclopedia records orange, white, and black cats, and possibly long-haired ones. By the sixteenth century, a great variety of cat colors and patterns are evident in beautiful Renaissance paintings, not to mention some lovely sketches of solid-colored cats by Leonardo da Vinci. To the east and a thousand years ago, cats of many different colors and patterns are present in Chinese paintings.

The best evidence of medieval feline diversity, however, comes from

Thailand (formerly Siam). A volume, the *Tamra Maew* (*Treatises on Cats,* or in its full form, *Treatises to Tell Cats' Characteristics*), passed down and revised through the ages, is thought to originally have been penned in the fourteenth century.

Lavishly illustrated and with text comprised of Thai poetry, the book is essentially a breeder's guide and illustrates a variety of cats. All are long and slender, with a thin tail and a long muzzle. The cats include an off-white puss with dark extremities (ears, legs, and tail—"points" in catworld), an obvious forebear of today's Siamese cats; a gray cat; a white cat; a gorgeous coppery-brown cat; and a black cat. Most of the remaining varieties are differently patterned black-and-white cats; subsequent editions of the *Tamra Maew* are inconsistent in describing them, though there are always seventeen different types.

Taken together, these depictions indicate that over the course of a millennium and a half, *Felis catus* around the world transformed from a wildcat-looking mackerel tabby to a species exhibiting a riot of colors and patterns. This transformation begs the questions of where the variation came from and why housecats are so much more diverse than their ancestors and other wild feline species. Let's focus first on the source of the variation.

We've already discussed one route. Hybridization can import new variants from one species to another. In fact, to a limited extent, that may have occurred with the domestic cat. The European wildcat is more darkly colored, as well as stockier and with a rounder head, than its African relatives. As *Felis catus,* sporting its African wildcat attire, spread through Europe, it undoubtedly intermingled with the native Europeans and took up some of their DNA. Very likely, that's why many domestic cats today, like our Henry, have that gray Euro-wildcat look.

But hybridization is only a small part of the story for a simple reason: most of the variation in domestic cats today doesn't occur in any wildcat population (or didn't until they started interbreeding with domestic cats). Housecats didn't obtain their orange, black, or white colors, much less tortoiseshell, van, or seal-point patterns, by mingling with the local wildcats, because historically the limited wildcat palette didn't include those colors or patterns.

The other source of variation is mutation, a change in the DNA. Some mutations have no effect whatsoever, but others cause a change in some aspect of anatomy, physiology, or behavior. For example, a single mutation in one gene can cause a cat that otherwise would have been mackerel striped to instead possess a swirly, almost psychedelic pattern on his sides seemingly straight out of van Gogh's *Starry Night* (such cats are referred to as "blotched" or "classic" tabbies).

A blotched tabby.

Another mutation can produce orange color; mutations in several different genes can result in a long-haired cat.

Does this mean that if we sequence the genome of a cat, we will know what he looks like? Not quite yet, but a lot of progress has been made. In some cases—such as blotched tabbies and long hair—geneticists have discovered the specific DNA changes responsible for these traits. For many traits, though, we're not there yet. For example, we know by analyzing inheritance in a pedigree that orange color is caused by an allele at a single gene (I'll provide some examples of how this is done in chapter fifteen), but the gene itself has not yet been identified.

Mutations are much more common than people realize, in humans as well as felines. Now that it is possible to sequence the entire genomes of individual cats, the mutation rate can be accurately calculated. By comparing the genomes of an individual and his two parents, scientists can identify DNA base pairs of an individual that are different from the DNA of both parents. Such differences are the result of the DNA mutating.

Using this approach, recent studies have calculated that the average housecat possesses forty-three mutations, that is, forty-three differences in DNA compared to the DNA of his parents. This number is somewhat lower than the average number of mutations we humans possess, because the human mutation rate is forty percent higher and because we have a larger genome than domestic cats.

The great variety of modern housecats—the colors, patterns, hair length, and other features not seen in African wildcats—is the result of mutations that arose within the last two thousand years. But why don't those features occur in wildcats? Is it because the mutations never occurred in wildcats or because they did occur but failed to persist?

Wildcats have been around for hundreds of thousands, perhaps millions, of years and have undoubtedly experienced countless mutations, many of them similar to those present in domestic cats today. It seems very unlikely that wildcat populations lack the variation seen in domestic cats because the requisite mutations never occurred; rather, it's much more likely that many of these mutations appeared in wildcats at some point but failed to become established.* The reason for the impermanence of these mutations is natural selection.

The mackerel tabby pattern provides excellent camouflage for a wild cat trying to blend into scrubby savanna or dark forest vegetation. Imagine a population of African wildcats living in, say, Sudan. Suppose a kitten is born with a mutation that makes him orange or white, or maybe a calico. Lacking camouflage, his prey would see him coming at a great distance. Just as importantly, his predators—of which wildcats have many—would also easily detect him. He wouldn't last long, and with his demise, the mutation would disappear from the gene pool.

This explanation seems very likely, but surprisingly, no one has tested the hypothesis. It wouldn't be that hard to do, at least in principle. I've conducted similar studies on lizards. Here's how you do it: you go to a population of feral cats living in the wild, getting by on their own and not being provisioned by people. Presumably, such populations, descended from strays, would contain a multitude of different colors and patterns. The hypothesis to be tested is that mackerel tabbies survive better and have more offspring than cats with different colors and patterns.

To test this hypothesis, you'll need to get to know as many cats as you can. One way to do that is to take pictures and learn to distinguish them by their patterns, whisker arrangements, or other markings, like Vered Mirmovitch did with the cats in Nachlaot. Alternatively, catch as many cats as you can and give each one a permanent identification

*Though keep in mind that the larger the population, the more mutations will occur. Today's 600 million domestic cats experience, in total, many more mutations than do the much smaller populations of existing wildcats. But wildcats have been around a lot longer, providing ample opportunity for mutations to have occurred.

marker, like the little chips you implant in your cat so that he can be identified if he's lost. While getting to know the cats, you also record each cat's coat color and pattern, and other information like sex, weight, and so on. Then you go home.

Six months or a year later, you come back and see which cats have survived and which ones haven't. One tricky aspect is knowing whether a cat you can't find is no longer alive or has simply moved somewhere else. You have to look far and wide, and hope that any undetected emigration is random with respect to coat color (that white cats, for example, aren't more likely to depart than cats with other colors). Then you get your records out and tabulate the results. Was survival related to coat color and pattern? Did a greater proportion of orange cats die compared to gray cats? Did striped cats persist better than calicos? If so, that's natural selection.

You could also catch kittens and take a DNA sample (say, from hair or saliva). Then parentage analyses could determine the mother and father of each cat, allowing you to test whether coat color or pattern was related to reproductive success.

Easy peasy, at least in theory. But as far as I'm aware, no one has ever measured natural selection in feral domestic cat populations.

Nonetheless, it's very plausible that mutations for a wide variety of colors and patterns have cropped up in wildcat populations over the eons, only to be quickly eliminated each time by natural selection.

But now let's think of early domestic cats, still looking like African wildcats, but no longer living in the wild. Sheltered from predators—most of which stay away from human habitations—and perhaps fed table scraps and so not needing to hunt, these cats would experience relaxed selection pressures. In this setting, being born orange, white, or tortie might not be disadvantageous.

Sometimes traits evolve not because they are beneficial, but just due to luck. The word "mutation" generally has a negative connotation, and indeed, some mutations are harmful. But not all. Many mutations don't matter one way or the other—in scientific terms, they are selectively neutral. Consider, for example, a mutation that affects whether you can roll your tongue. It seems unlikely that tongue-rolling ability has ever had much impact on an individual's survival or reproductive success. The same may be true for color and coat pattern in domestic cats living around people.

When a selectively neutral mutation occurs, by definition natural selection doesn't operate, so the fate of the mutation is determined by chance. Because a mutation starts out as uncommon—only one individual initially possesses it—most mutations eventually disappear from the population due to their failure to get passed on to the next generation. Occasionally, however, a new mutation hits the jackpot: just by serendipity, the bearer of the mutation lives a long, full life and has many offspring. If such good fortune continues for a few generations—remember, not because of natural selection, but just by luck—the mutant allele may become common in the population. In some cases, it may even replace the previous allele for that gene.

To a large extent, the diversity of cat appearance that we take for granted may be a product of such evolutionary randomness, a process called "genetic drift." And of course it's not just color. Many other odd traits can be surprisingly common in some populations, such as the possession of an extra toe (termed "polydactyly"*), kinks at the end of the tail, or no tail at all. It's hard to imagine why natural selection might favor these odd traits; their prevalence may be a result of genetic drift.

*Just like humans with extra fingers or toes, cats receive no benefit from having bonus digits.

On the other hand, perhaps we're not asking the right questions about why a trait might be favored by natural selection.

The most obvious source of positive selection for new traits among domestic animals, especially pets, is human preference. People like novelty. There's an evolutionary biology term for this—"negative frequency-dependent selection"—that is, traits are favored when they're rare. Imagine seeing an orange cat for the first time, or a black-and-white one. Pretty cool! Over the last two millennia, people may have favored individuals with new traits, feeding them more, taking better care of them, even intentionally breeding them, and this artificial selection may be responsible for the great variety of domestic cats today.* As we'll see in the next several chapters, such preference for novelty occurs even today, sometimes for traits that are rather bizarre.

Natural selection can be responsible for the diversity of domestic cats in another way. Many genes affect multiple traits. For example, cats with dark extremities (seal points) also have blue eyes. Because of such genetic correlations, one trait can become established in a population because selection favors another trait with which it is genetically correlated. Selection favoring seal-pointed cats will lead not only to more seal-pointed cats in subsequent generations, but also to more cats with blue eyes.

Orange cats are a possible case in point. They can be surprisingly common in some places. Why? Surely not for camouflage, except perhaps in a pumpkin patch. One possibility is that orangeness in cats is genetically correlated to some other trait that is advantageous. Indeed, several studies have shown that orange cats are heavier than

---

*Similar selection occurs in nature. Female guppies, for example, prefer to mate with males with unusual colors.

non-orange cats of the same age and sex. The researchers don't know what the cause is, but this correlation suggests that the mutation that causes orange color also affects some other aspect of the cats' biology, allowing them to grow to larger size. If being large is favored by natural selection, which is plausible, then the high incidence of gingers may be an indirect result of selection for large body size.

The advantage of orange color also may result from a genetic linkage with behavior. Tortoiseshell cats with a patchwork combination of orange and black (or similar colors, like yellow and gray) are widely known for their "tortitude." The website Meowingtons helpfully tells us that the term applies "to a cat with a tortoiseshell or calico coat that also happens to have a bit more, well, *cattitude*."* More helpfully, the website goes on to say that torties are "a bit more challenging, strong-willed, and can be possessive of their human" and also "fiercely independent, feisty and unpredictable."

It's possible to imagine that cats with such behavior would be favored by natural selection, and thus the commonness of orange color could be a consequence of this genetic correlation between orange color and feistiness. But is tortitude a real thing?

The only available data come from internet surveys. One poll of more than twelve hundred cat owners found that torties, as well as black-and-white and gray-and-white cats, were reported to be more aggressive to humans than cats of other colors. Another, in which participants were recruited through a Craigslist's community volunteer webpage, asked participants to rate cats of five different colors on ten

*With rare exceptions, only females exhibit tortoiseshell or calico patterns because orange color is a sex-linked trait in cats. As a result, males with the orange allele are solid orange, whereas females with the allele usually have a coat composed of orange and other-colored patches.

different personality traits. Somewhat confusingly, orange cats scored highest on being "friendly," whereas torties and calicos were the cats most often considered to be "intolerant," "aloof," and "stubborn."

Ideally, the relationship between temperament and color in cats could be studied more scientifically. All that would be required is a standard way of measuring temperament. Once the methods were established, scientists could either have cats brought to their lab or visit them at their homes, as Nicholas Nicastro and others have done to study vocalization behavior. But no one, to my knowledge, has taken this approach to test whether a link exists between behavior and coat color or pattern.

WE'VE COMPARED DOMESTIC CATS to wildcats for good reason: the former descended from the latter. But what about other cat species? Do their colors and patterns provide any information useful for understanding the variety of appearance in *Felis catus*?

For the most part, the answer is no—there are few parallels with other felines. Imagine a tuxedo tiger, a calico caracal, or a piebald puma. They don't exist. Indeed, other than the similarity of mackerel tabbies to the ancestral wildcat, and several recently developed breeds that were created to mimic spotted or striped wild species, almost no overlap exists between the color and pattern of domestic and wild felines.

With one exception. From the witch's sidekick to the Marvel superhero to Mowgli's friend and protector in *The Jungle Book*, black felines hold a special place in the human imagination.

And they're not limited to the household variety. Black panthers—which, in reality, are leopards—are the best known, but melanistic (all black) individuals are known to occur in fifteen cat species.* In mammals, melanism results from mutations that cause the pigment eumelanin to be deposited throughout the length of each hair, making it appear black. Eight different mutations have been identified that cause melanism in felines.†

Black domestic cats have a long and often unhappy history. Identified as witches' "familiars" and even thought to be sorceresses transformed into feline form, black cats were brutally slaughtered throughout the Middle Ages. Some hold that this massacre, which often extended to all cats, may have been responsible for population explosions of rats and, subsequently, the Black Death plagues resulting from fleas that traveled on the rodents' backs. Even today, black cats are considered by some to be an ill omen and are abused more frequently and adopted less often than non-black cats (kudos to my sister and the many others who adopt black cats for this reason).

Black cats are quite common in some domestic cat populations, yet they're absent from others. The relative rarity of black cats in parts of Europe today has been suggested to be the consequence of their persecution in the past. Though possible, this explanation lacks evidence.

In contrast, the distribution of melanism both within and between wild feline species is well understood. Ten of the fourteen wild feline

---

*Application of the term "black panther" to leopards notwithstanding, "panther" is most commonly used in reference to the species otherwise known as the mountain lion or puma, which occurs from Alaska to Patagonia in the Americas. Unlike many cat species, a black mountain lion has never been documented.

†An interesting sequel to this story is that melanism doesn't entirely eliminate the spots on leopards, jaguars, servals, and similarly adorned cats. Even though each hair is entirely black, there are different intensities of black; the slightly less black hair in some places produces a ghostly image of the normal spotted pattern when seen in just the right light.

species that exhibit melanism occur primarily in forests. This association also occurs among populations of wide-ranging species: in leopards, jaguars, and jaguarundis, melanism is much more common in forest-dwelling populations.

These observations suggest that being black may be an adaptation for enhanced camouflage in dark forest interiors, allowing melanistic individuals to be more active during the day. Scientists investigated this hypothesis by using camera traps to record activity patterns of melanistic and non-melanistic individuals of three forest-living species—the northern and southern tiger cats (aka, oncillas) and jaguars. The results were the same for all three: black cats are active in the day more often than spotted cats. And, as further evidence, black cats are also more active when the moon is full. The implication is clear: when there is some light, black color is advantageous in dark habitats presumably because black cats are harder to detect than spotted ones.*

If forest dwelling favors black color in other felines, it's a reasonable idea that it does so in feral domestic cats as well. Surprisingly, this hypothesis has never been explicitly tested, but there is a bit of evidence: in Australia, forty percent of the cats in a forested area in the southeast are black, compared to four percent in the desert Outback.

---

*If black color is advantageous, why aren't all individuals black? In a case like this, where one trait provides a survival advantage, we would expect natural selection to eliminate the other trait from the population. Scientists suggest there may be a counteradvantage to not being black but aren't sure what it is. One hypothesis is that many cat species have light-colored patches on various parts of their body, such as the back of their ears, which are used for communication. Many young cats, for example, use the ear markings to keep track of their mother when walking behind her. In black cats, those patches don't exist, potentially making it harder for cubs to stay with their mothers. Scientists are still investigating this and other ideas.

. . .

By living around us, cats moved into an ecological niche in which camouflage was unnecessary, thus setting the stage for diversification in color or pattern. But the role that humans have played in producing the great variety of domestic cats is uncertain. Perhaps colorful mutations proliferated simply due to genetic drift, one color being as good as another. Whether human preference was the primary factor promoting the evolution of these colors is unknown and probably unknowable.

But there's more to cat diversity than colors and patterns—the most interesting variation in *Felis catus* is in size, body shape, and hair texture. And there's no doubt that through artificial selection, our aesthetic preferences have played a major role in driving the evolution of this variety. But we can't claim all the credit. Cats have also adapted naturally to the environments in which they've lived for centuries without help from us.

Ten

A Shaggy Cat Story

In many mammal species, populations in colder climes have thicker pelage than their tropical counterparts. The explanation for this pattern is obvious: who wouldn't want a heavy coat when it's cold outside? The natural world is full of examples—think woolly mammoths! Wild felines show the same trend. Siberian tigers have long, thick coats, as do leopards and leopard cats from the Asian far north. In contrast, southern populations of all three species sport more lightweight outerwear. Wildcats, too, show the same trend; European wildcats from cold areas, in particular, are warmly clad.

So it makes perfect sense that as domestic cats moved north from their Middle Eastern origins, natural selection would have pushed them, too, to evolve longer, denser hair to adapt to chillier weather. In turn, we would expect breeds that people developed in the north to also be thickly furred as a result of being developed from already shaggy local cats. And that's exactly what we find.

I didn't know much about the Maine Coon* breed until Melissa and I started taking Nelson to cat shows. It was at our very first show, in Lenexa, Kansas, that I met the largest cat I've ever seen—Toby, a twenty-eight-pound Maine Coon riding around as a hood ornament on his breeder's motorized scooter. At first intimidated by his massive head and square-jawed, send-in-the-Marines look, I quickly learned why these long-haired gentle giants have become one of the most popular cat breeds (number two in 2021 rankings). It's not just their leonine good looks: friendly and laid back, these shaggy cats make excellent home companions and are particularly good around children.

A Maine Coon.

Maine Coons are an all-American original, the first cat breed to have originated in the United States at least a century and a half ago. Mainers put on their own cat shows in the 1860s, and when national competitions emerged late in the century, the big fellows took home many ribbons. By that time, Maine Coons already looked a lot like they do today; early descriptions emphasized their large size, long coat, large ears, and amply furred tail.

In the lore typical of the cat fancy,† fabulous stories swirl around the breed's origin. One clearly ridiculous idea is that they're derived from the mating of a stray cat and a raccoon, hence not only their

*Officially, this breed is referred to as the Maine Coon Cat in some cat organizations.

†The term "fancy" dates to at least the nineteenth century and refers to "the appreciation, promotion, or breeding of pet or domestic animals."

name, but also their size and bushy tails. Only slightly less fanciful is the tale that the breed is derived from long-haired pets of Marie Antoinette, sent over from France in anticipation of her soon-to-follow escape that, as we all know, never happened.

In reality, how the ancestors of these cats got to Maine will probably never be known, but what ensued is accepted wisdom. As the Cat Fanciers' Association website states, "The Maine Coon cat evolved through nature's own breeding program developing characteristics by following a 'survival of the fittest' evolution. The characteristics all have a purpose or function. Maine Coon cats developed into sturdy, working cats suited to the harsh winters and varied seasons of the Northeast region." As for those characteristics, the heavy coat, comprised of surprisingly silky soft hairs, provides warmth, as do the wraparound tail, ear tassels, and dense hair sprouting from within their ears; the well-tufted paws help in walking on snow and ice; and a tad of oiliness in their fur provides a degree of waterproofing.

In other words, the basic contours of the Maine Coon were shaped by natural selection, favoring those cats best adapted to living outside in Maine's famously chilly weather. No doubt, the look of these cats today has been refined and altered by the whims of modern-day breeders, but the Maine Coon essence was laid down by Mother Nature.

Maine Coons don't have a feline monopoly on cold weather apparel, however. Three thousand miles away in equally frigid Scandinavia, similar conditions begat a similar feline: the Norwegian Forest Cat.

Cat fancy rookies can be flustered, as I was, by the prospect of telling a Maine Coon from a Wegie. They're so similar, in fact, that some people think that Norwegian Forest Cats were brought over to North America, perhaps by the Vikings, and were the ancestors of Maine Coons.

If you go to cat-show-judge training school, as I did, they'll set you straight. Far from ancestor and descendant, the doppelgängers evolved their shagginess in distinctive ways: their heavy coats are superficially similar but are layered differently, and the mutations responsible for long hair are not the same.* A trained eye can pick out other differences. Both have an imposing head, but the Wegie's is more triangular and perhaps a tad less formidable. The wider ears and the straight-line profile from forehead to nose tip also distinguish the Norwegian from the Mainer.

Like the Maine Coon, Wegies have a mythic backstory, supposedly descendants of rugged forest denizens known from Norse legend to have prowled local lands for centuries or even millennia. As with the Maine Coon, these Norsk skogkatts are thought to have evolved their thick coats and other traits to adapt to the cold conditions in which they lived.

It was only within the last century, however, that people took over from nature in sculpting the Norwegian Forest Cat. Organized breeding began in the 1930s and took off after World War II. The people who created the breed started with a conception of what typical domestic cats living in Norwegian forests look like, then scoured the backyards and farms of Norway for cats fitting the bill. The result is another breed whose features evolved by natural selection.

Yet a third well-insulated cat hails from frosty climes. Many reports come down through the centuries of long-haired cats in various parts of Russia and nearby Asian countries. Harrison Weir, the Englishman responsible for the creation of cat shows and the cat fancy,

---

*Although the mutations do occur in the same gene, *Fibroblast Growth Factor 5* (scientists italicize the names of genes). Mutations in *FGF5* are also responsible for long hair in many breeds of dogs, rabbits, hamsters, and other mammals but not woolly mammoths.

was given a Russian long-hair sometime in the latter half of the nineteenth century. He described the cat as being hefty, with woolly fur, short legs, a big mane, and large, tufted ears full of long hairs.

Sounds like the Maine Coon and Norwegian Forest Cat, right? And presumably those traits had evolved by natural selection in free-living Russian domestic cats for the same reason as in Maine and Norway—it's freezing up there!

But then the trail runs cold. Russian cats stopped appearing in cat shows or even being talked about. They were pretty much forgotten, at least in the West, even if they were still roaming the streets, farms, and forests of northern Russian.

Forgotten, that is, until the late 1980s, when Russian cat enthusiasts decided to establish a local breed of long-haired cats. To do so, the breeders came up with a set of requirements that the ideal member of their breed should display. They then scoured the streets of St. Petersburg, Moscow, and other cities to find cats reasonably close to these standards. These animals and other subsequently located individuals formed the foundation for a new take on an old cat, evocatively named the Siberian and claimed to be descended from the fabled long-haired Russian cats of yore.

The traits that define the Siberian are said to be similar to those that characterize contemporary rural cats occurring in Siberia, in northwestern Russia, and outside Moscow. The long, thick, triple-layered fur of these cats, in particular, distinguishes them from Maine Coons and Norwegian Forest Cats. This pelage makes the Siberian arguably the best-adapted cold-weather breed thanks to plentiful hair on the back, around the eyes and ears, and on the chest and belly.

One point to keep in mind is that nobody is claiming that these three breeds today are identical to their forebears that lived centuries ago. Quite the contrary, artificial selection has changed their appearance even in the last few years. Indeed, not only have the breeds diverged from their ancestors, but members of the same breed can differ among cat organizations due to different ideas about what the cats should look like. Maine Coons, for example, differ between cats registered in different American organizations, as well as between those in the US and in Russia. Siberians also differ between the two countries. These recent changes notwithstanding, the major features of these breeds reflect the action of natural selection, adapting them to survive in local conditions.

Long-haired northern cats aren't the only ones that bear the imprint of natural selection. Consider how wildcats differ. The African wildcat is tall and lanky, its head somewhat triangular and its fur short. These features make sense in the hot climates these cats occupy. Who wants to pack a lot of meat on her bones, much less wear a shaggy coat, when it's stinking hot outside? Conversely, the European wildcat is thickset, round-headed, and wears a longer coat.

These differences are paralleled in domestic cats around the world today. Cats on the streets of Thailand tend to be long-legged and slender, with wedge-shaped heads. The same is true for the cats of Cairo. Conversely, British street cats tend to be stockier and their heads rounder than their equatorial counterparts.

This geographic parallel in body conformation between wildcats and domestics suggests the following evolutionary progression. The ancestral North African wildcat was well adapted to living in hot places. As newly arisen domestic cats in Egypt or thereabouts headed east and south to equally hot places, natural selection maintained this physique. Meanwhile, as cats pushed into northern Europe and Asia, and eventually were transported to northern North America, natural selection transformed slender ancestral cats into a shorter, stockier frame. This process may have been aided by hybridization with European wildcats, transferring the genes for cold-adapted traits into the domestic cat lineage.

Cat breeds can be arranged on this same body-type spectrum. At one end are heavyset breeds with compact bodies and short, thick legs; their heads are round and their faces short.

A cobby cat.

As you'd expect, almost all such "cobby" breeds—such as British Shorthairs, Manx, and Selkirk Rex—have European roots. Persian cats, the most extreme example of this type of cat, are seemingly a geographic anomaly. As we'll see in chapter fifteen, however, names can be deceiving.

Breeds developed in warmer, more southerly areas are the inverse of the stocky northerners. These cats tend to be slender and long-bodied, with tall, fine-boned legs and a triangular, wedge-shaped head. Traditionally, these breeds have been referred to as "Oriental" or "foreign," but those terms have fallen from favor in other contexts and, in any case, are not descriptive, so I'll refer to these breeds as "slender." Just as with the longhairs, we can see the signature of natural selection in these breeds, but artificial selection has changed them from their ancestral form, some much more than others.

One recently developed breed, however, does retain the ancestral look, the Sokoke Forest Cat. What, never heard of it? Until recently, neither had I. It's a very rare breed, derived from a feral population of domestic cats from coastal Kenya.

Forty years ago, people living in the area noticed the local cats and liked their looks. So they captured a few of them and started breeding them with each other, maintaining their appearance. As such, the SFC continues to look very much like its wild ancestors. And guess what the wild domestic cats from this hot part of the world look like: tall and slender with triangular heads and short hair. The Sokoke Forest cat gives us a look at the very early stages of breed development, when a breed has just been created from the cats of a particular region and artificial selection hasn't yet changed it.

I WANT TO FINISH this chapter with two caveats. First, similarity between cats in a place today and those that lived in the same place in the past does not prove a genealogical link between them. An example of a supposedly old cat with recent roots is the Egyptian Mau. These slender, spotted cats are often portrayed as "descendants of the cats of the Pharaohs." However, the breed was created in the mid-1950s when spotted cats with a slender build were plucked from the streets of Cairo. These pusses may have looked somewhat like those painted on the walls of the pharaoh's tombs, but that's pretty thin evidence to establish a three-thousand-year link to the cats of ancient Egypt. In a similarly tenuous way, the Khao Manee, a breed of beautiful white cats recently developed from cats in Thailand, is said to represent, and perhaps be descended from, cats pictured in the *Tamra Maew*. Some

Russian scientists express similar skepticism about the supposed link between today's Siberians and the Russian longhairs of history and legend.

Second, despite what I've written in this chapter, I need to conclude by emphasizing that an evolutionary history of adaptation to local climatic conditions is not immutable destiny. Human-directed artificial selection can easily override the ancestral blueprint, causing the appearance and body shape of a breed to change, sometimes substantially, through time. Once a cobby cat does not mean always a cobby cat!

The American Burmese* is an example. Despite being an Asian-derived breed, American Burmese buck the geographic rule and are very compact, cobby, round-headed cats. How did this happen? American breeders decided their cats were too similar to Siamese cats and relatively quickly created a very different-looking breed through artificial selection, in the process eliminating the body conformation features their ancestors had evolved by natural selection.

---

*Not to be confused with the European Burmese, the breed to which Nelson belongs. What we in the United States call a European Burmese is referred to in the rest of the world as a Burmese, whereas what we call a Burmese is elsewhere called the American or Contemporary Burmese. Confusing! The two breeds diverged in the middle of the last century.

Eleven

Not Your Father's Cat

Mention a cat show and most people think of the Westminster Kennel Club Dog Show: smartly dressed trainers parading their beautifully coiffed and perfectly behaved charges around the ring; madcap agility trials in which speedy canines zip through challenging obstacle courses with nary a misstep. A feline equivalent is unthinkable.

And yet, cat shows do exist. I know, because I've attended many of them, both as a spectator and as a participant with Nelson. Cat shows are simpler than dog shows. There is no cat promenade and the competitors in the agility competitions (which are a relatively recent addition) generally lack the single-minded zeal of their canine counterparts.

Nevertheless, cat shows are still a spectacle. Imagine two hundred, or even eight hundred, yowling, purring, and snoozing cats packed into a show hall, showcasing the variety of the modern cat. The venues range from shabby high school gymnasia and bare-bones veterans' halls to hotel banquet rooms and large show halls. The rooms are filled with rows of long tables, jam-packed with colorful kitty condos; the competitors lounge inside their fabric walls, waiting to be called to the judging tables. Siamese cats yowl incessantly. Occasional shouts of "cat out" or "cat on the ground" lead to a few moments of excitement until the wayward puss is retrieved.

In the morning, the only people present are the cats' minders (called "exhibitors"), getting their charges settled into their condos and primping them for what's to come. As the day wears on, though, the hall fills with spectators, a diverse cross section of the cat-loving public who've plunked down three bucks to see the show cat spectacle.

If you remember the zany characters in the dog-show mockumentary *Best in Show*, you'll be disappointed to discover that the exhibitors are just ordinary folk with a passion for cats and a willingness to let their lives revolve around driving—or sometimes flying—to events weekend after weekend throughout much of the year. Like any group that gets together frequently to compete and socialize, there are deep friendships, intense rivalries, gossip, complaints about the judging, and all sorts of hijinks.*

Fascinating as the people at cat shows are, let's focus on the main event: the cats! The contestants on display are mostly refined and elegant; it's hard to beat a Siamese for savoir faire or a Norwegian for

---

*If you really want to get a sense of what cat shows and the people who bring cats to them are like, check out the 2018 Canadian documentary *Catwalk: Tales from the Cat Show Circuit*.

reserved dignity. Some will charm you with their looks or manner; you'll be surprised at the unexpected features of others. But above all, what these events display is the amazing variety of catdom. The long, sinuous fluidity of the Oriental, the regal majesty of a Maine Coon, the pantherine sleekness of an Abyssinian. Fluffball Himalayans. Pixie-faced Devon Rexes.

Cat shows reveal that *Felis catus* is not one cat, but many diverse brands of feline. And the cat cornucopia is growing rapidly. Breeders have capitalized on naturally occurring mutations to develop new breeds unlike anything previously imagined, including the curly-haired Devon Rex and the Ragdoll, named for its penchant for going limp when picked up. Some enthusiasts are looking in a different direction for new sources of variation, mating domestic cats with other feline species to produce the gorgeous spotted Bengal, the long-legged Savannah, and others.

The International Cat Association now recognizes seventy-three breeds of cats, and the number is increasing rapidly. All these breeds share an essential catness, yet they are becoming increasingly different, in many respects more diverse than the forty-two wild species in the cat family. How far can cat breeders push the boundaries of modern felinity? Does cat evolution have no limits?

To find out, let's go to Cleveland.

With more than a million square feet of floor space—enough to hold a squadron of airplanes with plenty of room to spare—the International Exposition Center is one of the largest exhibition venues in the United States. At one time a tank factory, the building has hosted boat, auto, and RV shows, presidential rallies, NFL Fan Fests, and trade shows and conventions. For many years, an amusement park featured the world's largest indoor Ferris wheel.

But I'm not at the I-X Center (as Clevelanders call it) for any of that—I'm here to see cats! Once a year, the Cat Fanciers' Association hosts its International Cat Show, the largest gathering of its kind in the United States.* And in 2018 and 2019, that show was held in Cleveland.

I've been to my share of cat shows, but this one is like no other. Size, of course—most shows are held in spaces a fraction of the size of the I-X. And number of competitors—eight hundred cats is about five times the typical enrollment.

The International is much more than an enormous congregation of cats in a cavernous space. This is the Big Show, the World Series of Cats. Enormous banners hang horizontally from the rafters, each with a winsome photo of a different breed. "Siamese, Siberian, Somali, Sphynx, Tonkinese, Turkish Angora," the flags proclaim, and many more.

I'd never seen a Khao Manee, the beautiful white cat of Thailand, but a quick scan of the banners directs me to the back of the hall on the left, where several reside in their kitty condos, waiting to be called to a judging station. Want to see a Singapura? These tiny, big-eyed cats are on the right, toward the front.

Cat celebrities are plentiful. Sunglass Cat, with spectacles worthy of Elton John (worn because she was born without eyelids), meets any of the eleven thousand spectators willing to stand in a long line at the meet and greet. Sauerkraut Kitty models her latest outfits. Sphynx-wrinkly Sarah Pawcett looks surprisingly good in a Marie Antoinette costume, hairless no more thanks to a powdered wig. Friendly Ambassador Cats, such as Socrates, a bow-tie-bedecked Abyssinian, walk around on leashes, accepting pats and nuzzling the visitors.

---

*The CFA is the largest cat group in the world, with two million pedigreed cats in its registry since its founding in 1906. Their international show is perhaps the second largest in the world. I'm only aware of one that is bigger, the World Show put on yearly by a European group, Fédération Internationale Féline, which attracts as many as sixteen hundred entrants.

Entertainment abounds. A stage at the front left packs in hundreds of spectators for performances by the Savitsky Cats, showing off the amazing tricks that wowed Simon Cowell on *America's Got Talent*. In the agility ring, cats sometimes zip quickly into and over a variety of obstacles, though just as often they dawdle, to the tenders' frustration and the viewers' amusement, getting a good sniff of every item on the course. Lectures on cat origins and behavior are presented regularly, and an "adopt-a-thon" finds homes for more than a hundred cats during the two-day event.

But the show cats, all eight hundred of them, are the main event. The CFA recognizes forty-five breeds, and all but three are represented at the show.* The variety of these breeds is astonishing.

CAT SHOWS ARE A SPECTACLE in many ways. But to a biologist, the extravagance has a special significance, highlighting the power of selection to produce an enormous amount of evolutionary change over a short period of time. We've already seen what natural selection can do; cat shows allow us to explore the outcome of the cat fancy's artificial selection.

Let's start with their outerwear. There are more breeds of short-haired cats, but due to the immense popularity of a few breeds—Persians, Maine Coons, and Ragdolls—there are often more long-haired competitors at a show.

Breeds differ in the colors and patterns on display. Individuals of a few breeds, such as the Russian Blue and Havana Brown, are all the

*Missing are three rare breeds: the American Wirehair, LaPerm, and Turkish Van. Other cat organizations are more liberal in their approach and include many breeds not certified by the CFA—Highlander, Snowshoe, Serengeti, Savannah, Australian Mist, Chausie, Kurilian Bobtail, and Donskoy, to name a few.

same color (by the way, in cat and dog circles, the color gray is referred to as "blue" for reasons I don't understand); other breeds come in several options (for example, Egyptian Maus can be silver, bronze, or smoke); and for some, just about anything goes (Maine Coons, Manx, and Japanese Bobtails, among others).

Coat patterns vary similarly. For example, tabby stripes are common in some breeds, such as Maine Coons, Orientals, and Manx, but are absent in other breeds like Siamese and Chartreux. A similar situation exists for calicos and other patterns.

This variety of hair length, color, and pattern occurs among non-pedigreed cats as well. But other hair attributes are found almost exclusively in pedigreed cats. For example, a number of breeds have wavy, curly, or wiry hair (Devon Rex, Cornish Rex, Selkirk Rex, LaPerm, and American Wirehair, to name a few). Sphynx and several other breeds appear to be hairless, though, in reality, they often are covered in a coat of fine down. Taking this one step further, or maybe only half a step as far, the Lykoi's hair mostly falls out twice a year, the rest of the year occurring in patches, described on the breed's official Cat Fanciers' Association webpage as "sparse . . . hair on legs, feet and face mask which gives the appearance of a werewolf." And they're not kidding—I wouldn't want to run into one of those grimalkins in a dark alley.

Now, on to the head, which has been remarkably transformed in some breeds. Take a look at a modern-day, award-winning Persian and you'll notice something is missing: the nose! In its place are two little nostrils situated directly between the eyes (that's right, between the eyes, not below them like an ordinary cat, or you and me). Look from the side and you'll see the cat's face is a straight line from the eyes to the chin. Think I'm talking about a few abnormal cats? Here's how the official breed standard describes the ideal Persian: "When viewed

Persian cat from the front and in profile.

in profile . . . the forehead, nose, and chin appear to be in vertical alignment."

And it's not just the head. The rest of the Persian has been similarly compressed, producing a very short-bodied, compact, heavy-boned cat. It's as if someone put a normal cat into a vise and squeezed on both ends.

You may well wonder whether the abnormal nasal configuration is bad for the cats. Indeed, it may be—we'll discuss this and some other problematic cat breed traits in chapter fourteen.

At the same time breeders were compressing Persians and eliminating their noses, they were doing the exact opposite to Siamese. The typical Siamese cat in the mid-1960s was a slender feline with a somewhat wedge-shaped, triangular head. In just a few decades, breeders have transformed that cat into something otherwise unknown in the animal kingdom.

Today's Siamese has a long, narrow, and pointed face. Counterbalancing this arrowhead are large, widely spaced ears. The result is that the head of modern Siamese cats has a striking triangular symmetry.

The rest of the cat has been similarly elongated. The torso is cylindrical, the legs long and dainty, the tail wispy. These "slinky cats" are the exact opposite of Persians in every dimension. In the name of com-

A modern Siamese cat.

plete transparency, I will admit that I find today's Siamese to have a refined elegance—they may be unusual looking as cats, but they do have eye appeal.

Anyone who doubts whether evolution can occur need look no further than the Siamese and the Persian. The differences among breeds of cats and other domesticated plants and animals, produced by a few decades of managed breeding, provide strong evidence of the power of selection to rapidly alter the anatomy and behavior of a species.

Indeed, thanks to selective breeding, modern Siamese and Persian cats are unlike any feline species that have ever existed, either today or in the past. They are more different from each other than a lion is from either a cheetah or a domestic cat. If these breeds didn't exist and paleontologists found a fossilized Persian cat, they would publish a paper in a leading scientific journal describing the fossil not only as a new species, but placing it in a different genus, perhaps even a different subfamily of the Felidae.

I realize this is a pretty bold claim. But think about the Persian cat and its nasal deficiency. Now compare that to lions and tigers and ocelots and bobcats. Go to the internet for a refresher and look up some of the more obscure wild species. Guess what: they all have well-formed noses. Some have longer faces than others, but they all have distinct muzzles. And their nostrils are below the eyes, not between them.

Siamese are less extreme, but again, there are no other cats alive today, nor any fossil felines ever discovered, that have such a long, narrow, and tapering muzzle. Nor do any cat species have comparably big ears. In other words, both breeds exhibit facial structure more extreme than anything natural feline evolution has produced in the cat family's thirty-million-year existence.

The differences between modern-day Persians and Siamese capture most of the variation in head shape among cat breeds, but there are other dimensions: the massive square heads of Maine Coons; the impish, high-cheekboned Devon Rexes; and the Roman-nosed, egg-shaped heads of Cornish Rexes, for example.

Then there's the diversity of ears: tiny and bat-eared; high on the head and widely set; tufted and with a thick grove of hair growing from within (termed "ear furnishings"). And the odd attraction of folded-over ears: the pinnae of Scottish Folds bend forward so that the round-headed cat appears to have no ears at all, whereas those of the American Curl curve backward, sometimes touching the head.

THE SIAMESE-TO-PERSIAN CONTINUUM, embodying and exaggerating the differences between cobby European and slender Asian cats discussed in the last chapter, accounts for most of the variation in body form among cat breeds. As a general rule, long-legged cats have a slender body shape, whereas cobby cats tend to be short-legged. But a recently developed breed challenges this generality.

From the belly up, the Munchkin looks like an ordinary cat. Somewhat long torso, semi-triangular head, slightly on the slender side of the cat body spectrum. But then you look down. The cats are named after the people in *The Wizard of Oz* for a reason—their legs are extremely short. I am unaware of any precise measurements, but my estimate is that the legs of Munchkins are less than half the length of a typical cat's. Or to put it in simpler terms: Munchkins are the Corgis of the cat world.

A Munchkin.

There's not as much information available about the Munchkin as you might expect. When the breed first appeared, a few cat breeders were outraged, referring to the cats as "mutant sausages" and an "abomination." Beyond aesthetic considerations, there were concerns that the cats would be prone to the same spine and hip issues that plague Dachshunds, Corgis, and similar dog breeds. However, the skeletal structure of dogs and cats differ in important ways, and as far as is currently known, Munchkins are not afflicted with these maladies. There is some indication that the cats may be prone to several medical problems, but the evidence is inconclusive.

On the other hand, Munchkin enthusiasts have been making some dubious claims of their own. The International Cat Association's webpage starts: "The racy, low-slung munchkin is built for speed and agility." Elsewhere, others make the comparison even more explicit: race cars are low to the ground and really fast, so the same must be true for Munchkins. "The Munchkin's real strength is speed. They have an insane amount of energy and a knack for speed and agility, taking corners like a furry race car and staying low to the ground to get the most traction," says no less venerable a source than *Catster* magazine.

As a zoologist who has conducted research on the running abilities of animals, I must express my skepticism. Think about fast-running creatures, like cheetahs, gazelles, and greyhounds. Do they have stubby little legs? No, they have very long legs. The longer the legs, the more ground an animal can cover in a single stride. In my research, I built little racetracks to measure the sprinting capability of lizards, and the result was the same: longer-legged lizards are faster.

So the suggestion that Munchkins are the speedsters of the cat world strikes me as unlikely. Like so many other claims around these cats, however, there are no hard scientific data to consult. So to research the claim, I took the obvious route. I went to YouTube.

Sure enough, there are lots of videos of Munchkins scampering around living rooms. Let's make one thing clear at the outset—they are very cute, especially the kittens. But as for the claim that they're Olympian sprinters? I don't think so. They do show as much enthusiasm as any cat chasing a ball or a laser pointer, but I'm sure that in side-by-side trials, most cats would kick their butts down the track. But don't take my word for it—check out the videos for yourself.

Having said that, I will admit that it is conceivable that the shorter legs of Munchkins allow them to turn more rapidly. In part, that's because it's easier to execute a turn at lower speeds (think about driving), but also because, being lower to the ground, their center of gravity is easier to redirect.

I checked my intuition with Robbie Wilson, an Australian expert in the field of biomechanics. He agreed that short-legged animals should be able to turn more quickly and noted that "it's a very common observation that shorter-limbed soccer players have much tighter turning ability. I also have two Chihuahuas that have different limb lengths and love chasing each other around. The shorter-legged one gets chased down quickly, but then his turns are incredible. Great fun to watch."

Initially, some also claimed that Munchkins are unable to jump. That, too, is incorrect. Quite the contrary, they are avid jumpers, though chairs and coffee tables are more in their range than kitchen counters (this is one of the purported advantages of the breed). They also apparently will stand on their hindlegs, rabbit style, to get a good look around.

According to *Guinness World Records,* the shortest cat in the world is Lilieput, a nine-year-old Munchkin who stands all of five and a quarter inches at the shoulder. No wild species of feline has legs anywhere near as short as the Munchkin's, which suggests that the trait is not favored by natural selection. On the other hand, there are anecdotal reports of semi-feral populations in which short-legged cats are quite common. These wild Munchkins seem to be healthy and succeed in passing the trait on to the next generation, which indicates that detrimental health effects of the trait are not overwhelming.

A TAIL'S A TAIL, you might think, but you'd be so wrong. Cat tails vary from the whippy Cornish Rex's to the plumed splendor of the Turkish Van and the wraparound snuggie of the Maine Coon.

But this variation only applies to cats that have tails. Several breeds are characterized by having only a bobbed tail, a nubbin, or no tail at all. Some of these breeds are recent in origin, but the tail deficiencies of the Japanese Bobtail and the Manx go back centuries.

Indeed, in parts of Asia today, cats with kinked, bobbed, or otherwise unusual tails are quite common, comprising two thirds of the cat population in some areas. My beloved Siamese, Mauritia, was one such cat, with a kink at the very end of her tail, the last vertebra or two bent at a 90-degree angle to the rest of the tail. This trait used to be common in Siamese, but was mostly eliminated by breeders over the years.

Why these quirky tails are so common in Asia is not clear. Certainly, having a misshapen or absent tail is not advantageous—tails can be important for balance, warmth, and communication—so natural selection is not the explanation. It's always possible that tails of this sort are preferred in Asia for some reason, and thus selective breeding has promoted these traits. There are, in fact, some relevant legends, such as the story of a Thai princess who would remove her rings before bathing and place them on a palace cat's tail, the kinks in which kept the rings from falling off.

The other possibility is that kinked tails are selectively neutral. Perhaps some of the first cats to arrive in Asia, just by chance, had the genetic mutation causing this trait. Such a "founder event"—a type of genetic drift—would have led to a high frequency of malformed tails in the population, as long as they weren't detrimental (if they were, then natural selection would have removed the allele from the population). Possibly, such a random event spawned human preference for the trait, leading to preference for subsequent tail mutations when they arose. This, of course, is complete conjecture. For now, the tale of the malformed tail remains untold.

Twelve

Incessant Jibber-Jabber

So far, we've talked about differences in the physical characteristics of cats—their fur, their legs, their tails. But as anyone who's familiar with cats knows, they vary considerably in their temperament and behavior as well. And nowhere is this variation more evident than at a feline extravaganza.

The first thing you need to know about a cat show is that it is not one competition, but several.* Each show has multiple judges, usually between four and twelve, and each judge presides over an independent competition. That is, each judge examines every cat before choosing best in each breed and best in show, with kittens, neutered cats, and

*Cat shows vary somewhat in rules and procedures among cat organizations and in different countries. My descriptions are based on shows run by the Cat Fanciers' Association in the United States.

intact cats judged separately.* Depending on the number of judges, a show can have as many as twelve Best Cats in Show. The International Show is an exception in that the judges confer at the end to come up with an overall Best in Show.

Judges have their own "rings" where they examine all the cats. The ring consists of a table, sometimes with a cat scratching pole on it, surrounded on three sides by tables covered with wire cat cages. Throughout the day, the loudspeaker blares "Cats 102 through 114, please make your way to Ring Six," at which point the exhibitors remove the cats from their kitty condos, carry them to Ring Six, and place them in one of the holding cages.

The judge then goes through the cats one by one, removing each cat from her cage and placing her on the judging stand. The evaluation is thorough: the judge carefully inspects the cat's head, lifts the cat up to get a good look at her body conformation, wiggles a toy or pheasant feather to get a good look at her eyes, and so on. The appraisal varies, as the judge examines the traits considered important for each breed. Siamese, for example, are supposed to be long and slender, with cylindrical bodies. Judges will often pick these cats up—one hand in the armpit, the other on the belly in front of the hind legs—and stretch them out to assess whether they are appropriately tubular.

Most show cats have been doing this since they were kittens and are quite accustomed to being poked and prodded. They get on the judging stand, scratch at the pole, and chase the toy with abandon,

---

*Dog shows do not allow neutered animals to compete. Cat shows are more inclusive but keep the two groups separate because males mature differently after being neutered due to the lack of testosterone (becoming less muscular and possessing smaller heads) and because breeders often keep their best cats unaltered so that they can breed them and continue the lineage. Speaking of inclusiveness, cat shows now commonly have a category for non-pedigreed "household pets."

A judge examining a Siamese cat.

swatting and pouncing to their heart's content. A good judge will put on a great show, entertaining the spectators and exhibitors seated in front of the table.

The cats are judged on how closely they meet the breed standard*—in other words, how close to the ideal head, body, tail, and coat characteristics prescribed by each breed's ruling council.

For the most part, judges play with the cats to get a good look at them, rather than to evaluate their personality. That's how it's supposed to be, anyway. But judges have a lot of discretion in how they award points, and more than a few admitted to me that "unfriendlies" don't go very far. Indeed, a cat that bites a judge—which does happen, though uncommonly—is immediately disqualified. A cat that bites three judges is banned from competition forever.

The judging ring is a great place to get a handle on behavioral

---

*The term is shortened from its original phrasing in the nineteenth century, "the standard of points of excellence and beauty."

differences among the breeds. Judges learn to keep their hands on some breeds to keep them from jumping off the table, whereas cats of other breeds stay put without restraint. Persians, for example, are fairly placid and definitely stay four on the floor. If there's a scratching post on the judging table, as there often is, Cornish Rexes often climb it and teeter on the top. Some breeds, such as Korats, have a proclivity to steal cat toys from the hand of the judge and not give them back. Ocicats, European Burmese, Turkish Angoras, and Japanese Bobtails can all be very playful.

Cats are called to the judging ring in alphabetical order by breed, starting with the Abyssinians. Because the ring has fourteen or so cages, multiple breeds are present at any one time. And by alphabetical misfortune, that means that Nelson, my European Burmese, was usually placed near the Colorpoints.

I had never heard of this breed until I started showing Nelson. Colorpoints are Siamese cats that don't have the traditional, accepted Siamese color and patterns (essentially, bodies that are white, cream, fawn, or light steel-gray, with the famous Siamese dark points on the feet, tail, face, and ears). At some time in the past, Siamese fanciers wanted to diversify the Siamese palette and introduced a wide variety of colors and patterns. Traditionalists in the Siamese community would have none of it and refused to allow registration of such cats as Siamese. All-out warfare prevailed. Eventually, the solution was to designate another breed, the Colorpoint, for cats wearing untraditional colors (this actually occurred twice, producing yet another breed, the Oriental, again simply a Siamese in different clothes).* There's no need to get any deeper into the politics of the cat fancy (of

*Not all cat organizations recognize these breeds—especially the Colorpoint—as distinct from Siamese.

which, I've learned, there's quite a lot). Suffice to say that there are three breeds of fairly identical-looking cats.

But back to Nelson's misfortune. Every time he was called to a judging ring, Melissa or I would walk him to the ring, place him in a cage, then take a seat nearby. And then for the next fifteen minutes, we would be serenaded by "mi-Ahhh, mi-Ahhh" as the Colorpoints caterwauled nonstop. They never shut up! Who could blame Nelson for being in a bad mood in the face of such incessant jibber-jabber?

Cats of the Siamese clan are famous for their loquaciousness. "Constant nonstop talkers," says one website devoted to the breed. "Their reputation for endless chatter is well-deserved," agrees another.

One research team surveyed eighty veterinarians who specialized in cats, asking them to rank the behavioral differences among fifteen different breeds. Of all the traits examined, vocalization was the one that differentiated the breeds to the greatest extent. And which breed was the most chatty? By a landslide, Siamese, followed by Orientals.* The least talkative, in case you're interested? Persians and Maine Coons.

The survey revealed that breeds differ in many other behavioral characteristics as well. Want an active cat? Bengals and Abyssinians would be the way to go. Want a snuggly layabout? Think Persian or Ragdoll. Bengals and Abys are the most playful, along with Tonkinese and Siamese; Persians and Sphynx are the least fun-loving. Ragdolls are tops in the lover category; Manx join Bengals and Abys at the bottom of the affectionate list.†

*Colorpoints were not included in that study, but a more inclusive ranking in a popular encyclopedia of cat breeds puts all three flavors of Siamese at the top end of the talkativeness spectrum.

†When I discussed these results with friends who live with pedigreed cats, there was much upset and consternation. All I can say is: don't shoot the messenger. I'm just reporting what the study said—decide for yourself whether you believe it.

Several research groups have conducted surveys of people living with cats, rather than of vets. Although results differ in some details from one study to the other, the general conclusions are very similar. Cat breeds differ in almost any behavioral trait you can think of, such as aggression toward other cats; aggression toward family members; shyness toward both strangers and novel objects; even in furniture scratching, toy fetching, litter box use, wool sucking, and spraying in the house.

As a scientist who has conducted behavioral research, I have to add a caveat: I wish researchers would conduct a study in which they directly observed and recorded the behavior of different breeds, rather than using people's opinions expressed in surveys. Although I suspect that the results of these surveys are fairly accurate, all kinds of biases can creep into survey data. For example, people may expect cats of a particular breed to behave in a certain way and may elicit such behavior by how they interact with their cats. Or people who own Ragdolls may, hypothetically, be the sort of people who are more generous in their scoring than, say, people who own Bengals—in other words, differences in scoring of the breeds may reflect the personalities of the people who prefer each breed, rather than in the breeds themselves. A properly designed research study would eliminate such biases.

BUT LET'S GET BACK to Nelson. What endeared him to us was his incredibly affectionate nature. On a scale of one to ten, where ten is the most affectionate, the *Encyclopedia of Cat Breeds* puts European Burmese at a nine, behind only, of all cats, the Sphynx.* And once

*A Finnish survey of 4,316 cats (well, of their human companions) also put them in second place, this time behind Siamese and well ahead of Sphynxes.

Nelson arrived at our house at four months of age, he proved that he was a nine-plus.

Before his arrival, I never understood what people saw in dogs. But now I get it—it's heartwarming to live with an animal companion who seems genuinely happy to see you and who clearly enjoys your company. Nelson follows us around and starts purring, loudly and pleasantly, immediately upon being picked up or even being looked at. He quickly rose to the rank of Best Cat in the World.

And that's why what happened next was so surprising. At first, Nelson did fabulously at cat shows. My basement office is festooned with the colorful ribbons he won: blue, orange, and white streamers for Sixth Best Kitten in Illinois; long blue for Third Best Kitten in Wichita; red-and-black for Second Best Premier in St. Louis; yellow-and-white for Best Premier at the International in Cleveland! Cats get points for every successful outcome; amass enough points and the cat gets promoted to the next level, such as Grand Premier or Grand Champion. Nelson seemed destined for greatness!

All that success even though he had a minor weight issue. Euro Burms are known to be surprisingly heavy for their moderately slender build, but we may have given him a treat or two too many. We knew there was a problem when a friendly judge, while awarding him a ribbon, cautioned us that he was becoming a "chunk" and would have to be renamed "Jabba, the European Burmese."

We fixed the weight problem, but then the bottom fell out. Nelson decided that he didn't like cat shows. We're not sure why: maybe it was the smell of all those cats, or maybe it was the incessant bleating of the Colorpoints. Melissa thinks the breaking point was at a show in Kansas, when an unaltered male sprayed all over a nearby kitty condo in a most odoriferous fashion.

Whatever the reason, our gentle, loving boy became an unhappy camper. His demeanor and body posture clearly indicated he was not pleased to be at the show. On the judging stand, he started growling, or sometimes even hissing at the judges. It seemed like only a matter of time before he bit one. His quest for Grand Premiership came to an end.

And then, in one of the few positive developments during the pandemic, the Cat Fanciers' Association created a new level of distinction, between Premier and Grand Premier. All that was required was that your cat had amassed an intermediate number of points (fewer than required to achieve grand premiership) and that you were willing to write a check for fifteen dollars. Bingo on both counts! Nelson is a winner after all! Henceforth, he is to be treated with the respect and deference that are rightly his, and to be referred to by his official cat-world name: Silver Premier Mayonaka's Nelson Losos.*

More importantly, now that he's transitioned into his post-showbiz retirement, he's resumed the easygoing, affectionate ways that distinguish European Burmese as among the friendliest of cat breeds.

Dogs are renowned for behavioral differences among breeds, the result of artificial selection to perform different tasks like herding, guarding, chasing, fighting, and tracking. The thought of cat breeds being selected in similar ways is comical—there are no working groups of cats. Rather, cat breeds have been developed almost entirely on physical differences; no cat breeds have been created to perform particular jobs or functions.

But that's not to say that selection on behavior hasn't occurred during a breed's history. Just like any other trait, behavioral variation

*Mayonaka is the name of the cattery where Nelson was born.

among individuals can be the target of selection. If that variation is the result of genetic differences, then evolution can occur. In fact, as we'll see in chapter fourteen, selection on behavioral temperament has been part of the process of creating several new breeds.

Persians are a case in point. Harrison Weir described these cats at the dawn of the cat fancy in 1889 as "less reliable as regards temper than short-haired cats . . . In some few cases I have found them to be of almost a savage disposition, biting and snapping more like a dog than a cat. . . . My attendant has been frequently wounded in our endeavour to examine the fur, dentition, etc., of the Angora, Persian, or Russian." Frances Simpson agreed in *The Book of the Cat* fourteen years later, stating that "Persian cats are not so amiable or so reliable in their temper, as the short-haired varieties. I am inclined to think, however, that they are more intelligent. . . . They are apparently as keen hunters of prey as the short-haired cats."

These descriptions paint an image of Persians far different from today's placid layabouts. Though documentation is scant, this transformation seems to have been brought about by artificial selection, breeders choosing the calmest, least reactive cats to produce the next generation. Such equability may have been necessary because Persians, with their extremely long hair, require extensive daily grooming.

On the other hand, the behavior of a breed may have been inherited from the ancestral cats of that breed, rather than as a result of selection during breed development. Perhaps Abyssinians were initially derived from very active cats, for example. Almost surely that is the explanation for the high-energy antics of Bengals and Savannahs, for reasons we'll get to in chapter fourteen.

And then there's the talkativeness of Siamese. Is this a trait that anyone in their right mind would have selected for? Shortly after Siamese arrived in the West, they were already known to be chatterboxes.

"The breed is certainly the noisiest . . . of all the cats," as one breeder put it at the turn of the twentieth century. "They meow loudly and constantly, as if trying to talk, and to a deaf person at that," concurred another observer.

So, if selection was favoring this vocality, it must have occurred back in the breed's native land. Remember that the *Tamra Maew* was essentially a breeder's manual, so that is certainly possible. But the text has little to say about their cries, and Siamese cats in Thailand today (known as Wichienmaat) are said to vary in how garrulous they are and to be no more conversational than other local cats. As a result, there's little evidence that Siamese cats were ever selected for chattiness. One possibility is that even if all Siamese in Thailand weren't very vocal, the most loquacious cats may have been the ones sent to the West; this founder event—either intentional or accidental—could have led to the talkative breed we know today.

In many cases, how a breed was developed is lost in the mists of time. Even for those developed more recently, there are few records of behavioral selection, but that doesn't mean it didn't occur—indeed, many breeders emphasize that choosing cats that will make good human companions is an important consideration in the selection process. Still, it's hard to imagine breeders selecting cats because of their wool-sucking or toy-fetching behavior. Most likely, breed differences in these traits arose inadvertently either due to the idiosyncrasies of the individuals that founded a breed or because the behaviors are genetically linked to physical traits on which selection occurred.

Regardless of how these behavioral differences arose, their existence has two important implications. First, the behavior of members of a particular breed is reasonably predictable, and second, some of these behaviors affect what these cats are like as companions in the home. We'll return to these points in chapter fourteen.

Thirteen

Breeds Old and New

Cat shows and the cat fancy arose in the Victorian era in England, following closely on the heels of similar developments in the dog world. Most of the types of cats recognized in the late nineteenth century corresponded to differences in coat color, pattern, and length. Differences in body proportion or head shape received little attention, though Siamese had recently been introduced to the Western world, and their distinctiveness was noted. Most of today's breeds either did not exist or were very different from today's versions. Exceptions were the tailless Manx and the Maine Coon, news of which reached England just before the turn of the century. In other words, some variety already existed, but it was a far cry from our current diversity of cat sizes, shapes, and appearances.*

The transformation of two of the original feline standard bearers—the Persian and the Siamese—embody the changes that have occurred in felinedom. The November 1938 issue of *National Geographic* had an article, "The Panther of the Hearth," that featured many photos of prize-winning Persians and Siamese. With the exception of one snub-nosed Persian (called a "peke-faced" by analogy to the Pekingese breed of dogs), all the cats were beautiful, normal-looking cats. In fact, some of the prize-winning Persians and Siamese were fairly similar to each other in facial structure.

Today, as we've already discussed, Persians and Siamese are barely recognizable as members of the same species. How has this divergence been accomplished, and why?

Let's start with the *how.* At face value, the practice of selective breeding is straightforward. Imagine what happened when the first breeder decided to produce a more slender, triangular-headed Siamese. Variation is present in almost every population of any living species, including cats. So if you wanted to produce Siamese that are more slender, you'd simply survey the cats in the Siamese population, pick out the most slender ones, and let them mate with each other. It gets a little more complicated when you are trying to select for two traits simultaneously, because the slenderest cats don't necessarily have the most triangular heads, so compromises must be made.

In any case, you choose your animals, put them together, and hope

*A parallel trend is true for dogs. Although functionally derived breeds—the classic "working groups"—were developed over centuries to aid in various tasks, most dog variety is also a recent development.

they'll get it on. Then you look at all the kittens produced from these matings. Again, you do the same thing, pick the most extreme cats from that generation and pair them up. And so on, generation after generation.

Sometimes, it's three steps forward, two steps back. You get a cat with perfect eyes, but his other features aren't so good. So you breed him to get the alleles responsible for his eyes established in the gene pool, but spend several generations weeding out the alleles for his other features.

But selective breeding is only half the story. For selection to operate, variation must exist within a population. If, for example, all individuals have brown eyes, it's not possible to evolve some other color—not possible, that is, until individuals with some other eye color appear in the population either by mutation or immigration. Variation is as important to evolution as the selection that acts upon it.

And here we need to talk about genetics. In the rest of this book, I will primarily discuss traits that are determined mostly by a single gene. Traits like this usually occur in discretely different states: the ears curl backward or they don't; the tail is long or very short (or absent). These differences are the result of individuals having different alleles for a specific gene.

However, other traits are determined by the combined action of many different genes, each of which, on its own, has a relatively minor effect. Many of these are called "quantitative traits" because they don't occur in distinct alternative states, but rather represent a continuum, like body size or head shape (very narrow to very broad, and everything in between). For traits like this, offspring will tend to be the average of their parents, but this is just a tendency with a lot of variation. Mate a Siamese with a round-faced British Shorthair, for example, and most of the kittens would have intermediate-shape faces, but some would be more triangular and others would be rounder.

And finally, one last complication. The genetics of traits can lie between the two extremes just discussed. One or several genes can have a major effect on a trait, while many other genes have a more minor influence.

The continual production of new variation by mutation is key to understanding why evolution is a creative process capable of producing traits unlike any that existed in the ancestral population. Sometimes, a new trait will appear fully formed as a single mutation; we'll discuss examples of such traits shortly. But in other cases, selective breeding is more than just finding preexisting desirable traits and breeding individuals that possess them. Rather, new mutations occur generation after generation, incrementally modifying the traits that occur in the population, allowing evolution to build new traits that didn't previously exist. That's how you could start with cats in the 1930s that all had normal noses and end up with today's Persians. At first, breeders began with fairly normal-looking cats and selected the individuals with the shortest noses. Then, new mutations popped up producing individuals with even shorter noses.* By breeding those individuals, the average nose length decreased further. Then another mutation tweaked the nose even more. Mutation and selection, mutation and selection, generation after generation. Eventually, you end up with a cat without a nose.

Scientists have proven the effectiveness of this method in laboratory experiments over the last century. Fruit flies have been particu-

* Mutations also popped up producing individuals with longer noses, but breeders ignored those individuals. Mutations do not occur because they are needed; they occur randomly with respect to selection.

larly popular because they can be maintained in large colonies that contain a lot of variation and that produce numerous new mutations; their short generations (just a couple of weeks) also help by hastening evolutionary change. These experiments have shown that selection can rapidly change almost any feature of fruit flies—the size of their wings, the number of bristles on their abdomen, even their propensity to fly upward—in just a few years producing flies vastly different from those in the ancestral population.

This same technique, of course, is how farmers and ranchers have radically transformed corn, wheat, sheep, and cows from their ancestors to the domestic plants and animals we have today. And cats are no different: apply strong selection and they'll evolve very rapidly.

So it's easy to envision how breeders used standard artificial selection practices to transform Persians and Siamese from a normal cat appearance to the physiques they have today. But the bigger question is *why*. Just because breeders have the power to reshape a cat in odd and unconventional ways doesn't mean they have to do so. Putting on my amateur psychologist hat, I am puzzled by what drives someone to want to produce cats like these, especially the Persian. Who had the idea to take a beautiful cat with a normal face and get rid of the nose, leaving only nostrils between the eyes? Why did others go along with that plan? And why didn't the process stop at some intermediate point, rather than continuing until the Persian's nose was gone? The answers are probably buried in the notes and minutes of Persian cat fanciers' organizations, but I have been unable to locate them. So we'll have to leave the "Case of the Disappearing Nose" as an unsolved mystery. Instead, let's turn to the "Case of the Slinky Cat"!

Siamese came to the attention of the Western world in the 1870s, when several were exported to England and appeared in the first cat shows. The "Royal Cats of Siam," as they were called,* were an immediate sensation (though not always in a positive way; *Harper's Weekly* referred to them as "an unnatural, nightmare kind of cat").

By 1902, the Siamese Cat Club had been formed in England, devoted to promoting the development of the breed. The "Standard of Points" they produced includes:

"Shape—body rather long, legs proportionately slight.

"Head—rather long and pointed.

"General appearance— . . . a somewhat curious and striking looking cat, of medium size; if weighty, not showing bulk, as this would detract from the admired 'svelte' appearance. In type, in every particular, the reverse of the ideal short-haired domestic cat."

Recall that the common cat in Britain, especially at the turn of the last century, was a round-headed, heavy-boned, compact animal. The goal of these standards is thus to draw a clear distinction between Siamese cats and the familiar locals.

The 1903 book in which these standards were described—Frances Simpson's *The Book of the Cat*—is still a classic. The chapter on Siamese cats includes many photos of cats that look like the ordinary, "traditional" Siamese that used to be common, such as Tammy and Mauritia, the two Siamese I grew up with. The question is whether the club's new standards were simply a description of the status quo—the Siamese cats they already had in their homes—or whether they were a road map to the future development of a much more extreme physique. That is, when the standard writers said the head should be long and pointed, just how long and pointed did they have in mind?

*An appellation we now know to be incorrect; these cats were not the special province of royalty.

The photos in the 1938 *National Geographic* article indicate that for several decades, there was relatively little change in the Siamese breed. After World War II, however, the situation changed. In the 1950s, more extreme cats started appearing at cat shows. In the 1960s, breed standards were revised to explicitly favor these extreme forms. By the 1980s, cats with the traditional Siamese head and body conformation no longer appeared at cat shows—the slinky-cat look had become the norm.

The question is: why, after fifty-plus years, did Siamese breeders suddenly decide to transform the traditional type into the ultra-extreme cat of today? From discussions with cat authorities and by analogy to what happened in the development of dog breeds (about which there has been scholarly research), I suggest there are at least five possible explanations. Spoiler alert: I don't know the answer!

First, maybe this was a ploy by wealthy cat breeders to keep the hoi polloi out of the Siamese cat industry. By insisting on an extreme standard maintained by the cat fancy elite, perhaps the mom-and-pop breeders could be marginalized, leaving all the money to be made, not to mention the glory to be gained, to the select few.

Second, this is what cat breeders do. They have a vision of what the ideal cat should look like, and they consider it their job—their duty—to try to produce a cat that meets that ideal. A corollary of this view is that the current state of the cat breed never matches the ideal. It's a step in the right direction, but the breed always needs refinement. This explanation, of course, doesn't explain why it took fifty years for the transformation to gather momentum.

The answer may be, third, that all along, Siamese breeders were angling for the modern look, but until cats were born with the appropriate variation, selection couldn't occur. That is, if all the cats in a population have the same head shape, then there's no opportunity for selection, human-driven or otherwise, to produce evolutionary change. Maybe it took fifty years for the right mutations to appear.

Fourth, once a breed transformation begins, it's hard to put on the brakes. If the breed standard calls for a head that is long and pointed, then it's always possible to envision a cat whose head is even longer and pointier than the current state. Undoubtedly, this explains how the Persian ended up without a nose—who in their right mind would have suggested that as a goal a century ago, when Persians had normal muzzles and the idea of a slightly shorter one was conceived?

Fifth, extreme looks win prizes. Judging a cat show is not easy. Judges must be familiar with the standards for dozens of breeds. If the breed standard says the head should be long and pointed, then giving the first-place ribbon to the cat with the longest and pointiest head is easier than trying to remember the optimal degree of long and pointy for that breed. In this way, judges may drive continued exaggeration by continually favoring the extremes.

There is one amusing sideline to this story. Apparently, most people in the public—myself included—prefer the "traditional" Siamese look rather than the extreme new appearance now mandated in the Siamese breed standard. In the 1980s, some enthusiasts banded together to create a new breed for cats with the old look. This led to an enormous catfight, the members of the existing Siamese cat clubs once again battling tooth and nail to squelch the recognition of another breed. Nonetheless, they failed and the new (old) breed, the Thai, indistinguishable from Siamese of the past, is now recognized by some cat organizations alongside the modern Siamese.

The point of these various stories is that breeders have the ability to change a breed quickly and substantially. To a large extent, the changes are driven by the aesthetic whims of the people in charge of specifying the particular standards for a breed. You might well ask: "Just who are these people on the breed council and why do they get to decide the fate of a breed?" Different cat organizations have different rules about who gets to serve on their breed councils, but in all cases, the individuals are people who keep cats of that breed. Of course, that doesn't mean their opinions are inherently more informed or valuable than anyone else's. Nonetheless, those are the people who decide what the ideal cat of a particular breed should look like. As we've just seen with Siamese cats, their decisions can be out of line with those of the majority of people who care about a breed.*

THE PRECEDING DISCUSSION only concerns how breeds change through time. A hallmark of the cat fancy has been the enormous increase in the number of recognized breeds. Where do these new breeds come from? In many cases, it's just a matter of serendipity, someone coming across a cool-looking cat and thinking, "There should be a breed that looks like that."

One day in 1981, two kittens appeared on the parking pad in front of Grace and Joe Ruga's home in Lakewood, California. At the time, seven-months-pregnant Grace was, in her own words, "trying hard not to do anything" on account of a horrible, ongoing case of all-day-long morning sickness. Upon arriving home from work, Joe saw the kittens and interacted with them for fifteen minutes. He then went inside, told Grace of their presence, adding, "They look skinny. Don't feed 'em." Once Joe exited the room, Grace immediately got up and

*Because each cat organization has a breed council (or committee or other name), multiple standards exist for each breed. Usually, the different standards are similar, but not always.

brought them a dish of "some left over who-knows-what from the refrigerator" and a bowl of water. While doing so, she noticed they had "ears that curled back from their heads in a funny way."

Within a week, the kittens were members of the family with full indoor privileges. One of the cats disappeared a few weeks later, leaving the other, a beautiful black cat that Grace named Shulamith after the female protagonist in the Old Testament's Song of Solomon.

Shulamith, not only beautiful but also sweet and loving, had several litters, and the equally friendly kittens, some curly-eared, were given away. Everyone who saw them remarked on their unusual auricles, sometimes so backward curved that the tip pointed downward. Eventually, the Rugas realized that Shu's trademark ears were unknown in the cat fancy. A friend of a friend suggested the cats could be the foundation for creating a new curly-eared breed. So that's what they did. As for the name of the breed, many cutesy possibilities were suggested, but they went for short, descriptive, and patriotic: the American Curl.

An American Curl.

Once the Rugas decided to establish a new breed of curly-eared cats, they had a big decision to make: what did they want the American Curl to look like? After all, there's more to a cat than the shape of his ears. To get a breed officially recognized, the Rugas—with the help of a few others—needed to create a unique breed standard. And that standard had to include more than:

"EARS

"*Degree*: minimum 90 degree arc of curl, not to exceed 180 degrees. Firm cartilage from ear base to at least ⅓ of height.

"*Shape*: wide at base and open, curving back in smooth arc when viewed from front and rear. Tips rounded and flexible.

"*Size*: moderately large.

"*Placement*: erect, set equally on top and side of head.

"*Furnishings*: desirable."

Specifically, the standard had to include details about the shape of the head, the body conformation, the tail, and all the rest of the cat. Should the American Curl be solid and cobby like a Persian? Long and slinky like a Siamese? Rugged and powerful like a Maine Coon? Or maybe some novel combination of characteristics, something daring, new, and original?

The Rugas never considered any of these options. Shu was the foundation cat of the breed. American Curls should look like their original queen: somewhat slender with a moderately wedge-shaped head. Not too large. Perhaps with a few upgrades in ear size and chin strength.

Remember the definition of a "breed" from chapter one? A breed is simply a group of distinctive-looking domesticated animals whose offspring look like they do—they "breed true." The dog fancy has gone crazy in recent years, crossing almost any two breeds imaginable to produce the Saint Berdoodle, the Bullmatian, and many other so-called designer dog breeds. When Saint Berdoodles mate, if their offspring look like Saint Berdoodles, then that's a breed.* And if slender cats with a moderate wedge-shaped head and ears that curl backward tend to produce similar-looking kittens, then the American Curl's a breed, too.

Once the vision for the American Curl was in place, the Rugas faced

---

*For the most part, however, that's not how it works in dogworld these days. Rather, designer dogs are created anew each generation by mating parents of the two different breeds. The Australian labradoodle is a notable exception.

a second challenge: inbreeding. By necessity, all members of this new breed would be descendants of Shu—it's her mutation, passed on to descendants, that is the basis of the breed (unless, of course, a cat with the same mutation pops up somewhere else; possible, but it hasn't happened yet).

The problem with having a breed full of related individuals is that offspring of close relatives tend to be at risk for genetic diseases. The solution to this problem is to bring new, unrelated individuals into the gene pool to provide increased genetic variation.

The difficulty with this solution, of course, is that these unrelated cats do not have the allele for curled ears. Consider what happens with a dominant trait like curled ears. Suppose you breed a male American Curl that has two copies of the curled allele with an unrelated female cat. All the offspring will have one curled allele obtained from their father and one uncurled allele from their mother. Individuals that have two different alleles for a gene are called "heterozygous," whereas those that have two copies of the same allele are "homozygous." Because the trait is dominant, heterozygous individuals will have curled ears but will be carriers for the uncurled allele.

Now consider what will happen if two heterozygous American Curls mate. On average, one quarter of their kittens will inherit the uncurled allele from both parents, and thus will have straight ears. For a breeder, that's a bummer: you breed American Curls to produce cats with curled ears.

The straight-eared cats will not be allowed to breed, and thus the uncurled allele will slowly be winnowed out of the population. This weeding, however, takes a long time, because you can't tell whether a curly-eared cat is a carrier for the uncurled allele just by looking at him (genetic tests discussed in chapter fifteen can solve this problem once the gene for curled ears is discovered). That's the cost of bringing unrelated individuals lacking the trait into the breed—you end up spending a lot of time eliminating the traits you don't want.*

To obtain the unrelated, straight-eared cats to breed to their curled ears, the Rugas made life easy for themselves by selecting cats with body and head conformations similar to Shu's. To avoid trouble, they didn't want to use any pedigreed cats and risk, for example, the breeders of Turkish Angoras claiming that the Rugas' cats were just curly-eared Turkish Angoras (Shu was, in fact, more similar in appearance to a Turkish Angora than to any other breed).

So instead, they used regular old domestic cats, obtained, as Grace put it succinctly, "anywhere I could find them." Animal shelters, fast-food-restaurant parking lots, cat shows—if they had a body conformation like Shulamith's, they were good to go. Grace carried a cat carrier with her at all times, just in case.

As for color and pattern, they didn't care about that. "Because of their domestic origin, all colors and patterns accepted," says the

---

*If the trait of interest is recessive, such as the short tail of the Japanese Bobtail, then the situation unfolds differently. For a JBT to have a bobbed tail, a cat must have two copies of the bobbed-tail allele. Now consider mating that JBT with some random, normal-tailed cat. Such a cat would be very unlikely to be carrying the bobbed-tail allele, so it would be homozygous for the normal-tail allele. As a result, all offspring would be heterozygous and thus would have normal tails, because the trait is recessive. Think about that: you're a JBT breeder, and all your kittens don't have a bobbed tail. But you know your genetics and were expecting that. You then breed these cats to other, similarly produced, heterozygous cats. In this second generation, one quarter of the cats would inherit the bobbed-tail allele from both parents and thus would be homozygous and possess a bobbed tail. Once you produce enough of these homozygote recessives, your job would be done: by breeding these cats with each other, the normal-tailed allele would be removed from your breed. In other words, for recessive traits, the cost is up front: you produce a lot of cats early on not bearing the trait you want, but then you're rid of the alleles for the undesired trait; for dominant traits, in contrast, individuals keep on popping up with that trait for many generations as the allele is slowly removed from the population.

standard. This was America's cat and was meant to reflect the diversity of cats throughout the country. Hair length didn't matter either.

Five years after Shu showed up at Chez Ruga, the American Curl was a recognized breed.

THE SAME STORY, more or less, has occurred repeatedly over the last seventy years as new breeds have been developed from individuals with novel features. Here's just a sampler from cat breed history:

1950: A housecat living in Cornwall, England, gives birth to a litter of kittens, including one with tight curls making him look like a little lamb. Kallibunker is the first Cornish Rex.

1960: A stray female given shelter in a kindly woman's backyard gives birth to a curly-haired male. Kirlee becomes the progenitor of the Devon Rex.

1966: A Toronto housecat gives birth to a hairless kitten and the Sphynx is born.

1983: A woman finds a pregnant, short-legged cat hiding under a pickup truck in Louisiana and takes her home, where she gives birth to short-legged kittens. Voilà! The Munchkin.

2010: Two odd-looking, patchy-haired cats are born on a farm in Virginia. Concerned about their appearance, a cat rescue worker takes them to a vet. Someone else with an appointment at the same time sees the kittens and recognizes them as something special. Next thing you know, the Lykoi is an established breed.

Not to mention the Scottish Fold (1961), American Bobtail (1966), American Wirehair (1966), LaPerm (1982), Pixiebob (1986), Donskoy (1987), Selkirk Rex (1987), and Tennessee Rex (2004). And even this list isn't exhaustive!

The details vary from breed to breed, but the overall story is very much the same: a cat with a novel trait is born or wanders in off the street. Test breedings indicate that the trait has a genetic basis (as opposed to being the result of maternal exposure to some toxic chemical or some other non-genetic cause of a malformation). A breed standard is developed, usually based on the appearance of the foundation cat that first exhibited the mutation and making sure that the new breed isn't too similar to any preexisting breed. Other cats lacking the trait are brought in to boost genetic variation. After some time, the breed is established and recognized by one or more cat fancy federations (of course, some new types of cats are created informally and are never officially recognized).

Like the curved ear of the American Curl, mutations for some of the other traits have only occurred once (as far as anyone knows). A number of other traits, however, have appeared multiple times through the years. For example, in the four decades prior to their appearance in Louisiana, short-legged cats had been reported in England, Brooklyn, Stalingrad, Pennsylvania, and New England. Although in some of these places, the short-legged trait was passed on for several generations, it always eventually disappeared. After the establishment of the Munchkin, additional cats have popped up with the short-legged mutation elsewhere, and some of these cats have been incorporated into the breed.

In a similar manner, "bald," "naked," or "nude" cats—not to mention the "Mexican hairless"—have been repeatedly reported over the last two centuries. Mostly they, too, disappeared, but two different hairless mutations spawned the Sphynx and Donskoy breeds.

WE KNOW THAT THE short legs of the Munchkin are caused by an allele at a single gene. The hairlessness of the Sphynx is caused by an allele of another gene. Think about this for a minute, the possibilities . . . They didn't, did they?

They did.

Allow me to introduce the bambino, a buck-naked, wrinkly-skinned, short-legged freak of a cat. It's repellent. Or is it adorable? Regardless, you can't look away.

People have crossed Munchkins with many other breeds. Want a wavy-haired Munchkin? Just mate one with a Cornish Rex, but beware, not only is the hair curly, but the ears are almost as long as the legs. A Munchkin-American Curl cross? Yep, and I have to say, it's precious. A Siamese-Munchkin hybrid is a sublime mash-up of elegance and comedy. As for the Maine Coon-Munchkin cross . . . my, how the mighty have fallen.

The same craze has occurred with hairless cats. Seemingly every breed has been crossed with a Sphynx or Donskoy. Just try giving it a google.

This trend became so great that The International Cat Association approved a recommendation from its Genetics Committee that the organization "not accept any proposed breeds . . . that do not exhibit novel mutations. The current mutations would be reserved for currently recognized breeds exclusively." As one committee member noted, "This would end the seemingly endless applications for 'munchkinized' new breeds, and then deter the inevitable introduction of 'rexed,' 'Bob-tailed,' and 'Poly-ed' [i.e., extra-toed] everything else."

This decision highlights that just because a breed has been created doesn't mean that cat or dog registries need to recognize them. By the same token, many breeders are not members of these organizations, which means they are not constrained by the registries' breed standards. There is, in fact, a tension between those breeding cats for the show ring and those breeding them for profit. Show-ring breeders do sell their cats, but most do not cover their expenses. On the other hand, commercial breeders are in the business to make a buck. They are less concerned about meeting a registry's breed standard than in producing cats that are marketable to the general public.

THE EXAMPLES I've described here are pretty simple: a would-be breeder discovers a cat with a mutation for an unusual trait and builds a new breed around that trait. It wasn't rocket science to go one step further and come up with the idea of crossing two breeds, each defined by a single novel trait.

But some breeders have had bigger aspirations. And to realize them, they've taken an approach, unheard of in the dog world, that many would have thought impossible.

Fourteen

Spotted Housecats and the Call of the Wild

We've already seen that arranging a mating with a member of another breed is a fast way of getting new traits into a breed. But even with the great variation among cat breeds, there are limits to this approach: some traits aren't found in any breed.

But why not look a little further? Domestic cats can interbreed with wildcat populations; perhaps they can reproduce with other feline species as well?

And that's the approach some breeders decided to take a few decades ago. Their goal: to create a domestic cat with the look of a serval.

I have to digress for a moment. I've already mentioned servals several times, but I've managed to keep my emotions in check. But I can't maintain that composure any longer.

I've spent a lot of time in southern and eastern Africa, conducting research on lizards, giving lectures, leading nature tours, and just plain visiting. One of my most vivid memories is the day Melissa and I were driving through a game reserve in South Africa. Passing by a field of high grass, I suddenly noticed a pair of triangles sticking out of the grass. Could it be what I had been hoping to see for years? Raising my binoculars, I saw the triangles were the tips of two enormous ears. Underneath them was a tawny, spotted head sporting close-set eyes, a black nose, and a pointy muzzle. Yes, it was! To this day, Melissa makes fun of my uncontrolled scream-whisper, "It's a SERVAL!" The cat soon sauntered away, disappearing into the high vegetation, but not before the species had cemented its position as my all-time favorite feline species.

And deservedly so! Picture a small-cat version of the cheetah, only more beautiful, thanks to their vivid, black-spotted coat. Add those enormous upright ears, a stretch-Cadillac neck, and a banded half-length tail. There's no cat like it (see p. 58).

Serval hunting behavior is equally memorable. Those huge party hats have a purpose. Crouched, leaning forward, ears cocked downward, a serval can detect the slightest rustle of a small mammal in tall grass. Once locked in, the cat launches itself upward red-fox style, both paws landing simultaneously to pin the prey.

Servals have been tamed since the time of the ancient Egyptians, and zookeepers rate them as among the friendliest of wild cats. It's not rocket science to put two and two together and come up with the idea of creating a domestic cat version of the serval, and the plan is straightforward: arrange a love match between the two.

There's one big catch. If we look at the evolutionary tree of living feline species on page 73, we see that domestic cats and servals are on opposite sides. The branch leading to servals is an early offshoot, diverging from the branch that leads to domestic cats (and most other species) about ten and a half million years ago.

That's a long time. When species diverge evolutionarily, they become less capable of interbreeding. As years pass, different mutations pop up in the two lineages, and natural selection favors adaptation to their different environments. Consequently, the species become genetically distinct. This is potentially a problem because offspring get one copy of every gene from one parent and another copy from the other parent (except for sex-linked genes). If their parents' DNA is different enough, something tends to go wrong: the sperm may fail to fertilize the egg; the fertilized egg may fail to develop; the embryo may grow abnormally; the offspring, if born alive, may fail to thrive; and if she is physically fit, she may be sterile.

As a result, the ability to successfully reproduce decreases the longer two species are evolutionarily separated. Most species able to successfully interbreed have been evolving separately for a relatively short period of time, like the domestic cat and the European wildcat. Mammals as a group follow this trend: most species that are able to mate and produce living offspring have been evolving separately for less than four million years.

Bottom line: the long evolutionary divergence between servals and domestic cats suggests that it is unlikely that they could produce fertile offspring. A priori, developing a new domestic cat breed containing serval ancestry seems like a long shot.

DNA incompatibility is an issue, but a more prosaic problem turned out to be a bigger impediment. Servals are much larger than domestic cats, topping out at more than thirty pounds. When domestic tomcats mount female servals, they have all the right moves, getting a bite hold on the nape of the neck per feline protocol, but their "nether parts don't get to where they need to be," in the words of one expert. Imagine the poor tom's confusion and consternation. Sadly, the tryst is rarely consummated.

The opposite coupling has a different, more horrifying problem. Male servals sometimes bite too hard, injuring or even killing the poor female domestic cat. What breeders did to avoid that from happening, I don't know, but sometimes they succeeded, and lo and behold, half serval–half housecat kittens—the first members of the Savannah breed—were born.

A Savannah.

But could the nascent breed reproduce itself—were the offspring fertile? Many distantly related species can produce perfectly healthy, but sterile, offspring, like mules (the result of mating between a male donkey and a female horse). Obviously, you can't start a new breed if your offspring can't reproduce.

Quixotically, the female half serval–half cat offspring were fertile and the males weren't.* This is a common result of hybridization

*This reduced fertility confirms that servals and domestic cats are different species. The fact that serval-domestic cat hybrids have never been reported from the

between species, so common that it has a name, Haldane's rule, after the scientist who first noted it. In many species of mammals, insects, and other creatures, female hybrids are fertile and males aren't, whereas the opposite result almost never occurs. It seems to have something to do with the sex chromosomes; males of these species having an X and a Y sex chromosome, whereas females have two X's. Proof for this relationship comes from organisms like birds and butterflies, in which it is the female that has the heterogeneous sex chromosomes. In those species, infertile hybrids are almost always females. Scientists still aren't sure why this happens, but the pattern is clear.

Sterile males, fertile females. Is this the beginning of a new breed or the end of the line?

In this case, the Savannah glass is half full: fecund females are enough. Breeders mated their first-generation hybrid females (termed "F1s") with male domestic cats. F1s look a lot like servals but are substantially smaller, females averaging nineteen pounds. That's still a big cat, but apparently not so big that a tomcat can't get the job done in the bedroom.

Male F2s are still infertile, as are male F3s and F4s. But breeders kept on crossing their fertile females with male domestic cats and eventually—bingo!—in F5s, the males are fertile, at which point Savannahs can be bred with each other and further input from domestic cats is no longer needed.

There's a catch with this method. With every generation of backcrossing to domestic cats, the genetic contribution of the ancestral servals is reduced by fifty percent. A first-generation hybrid is fifty percent serval and looks it; an F2 is twenty-five percent serval, and so

---

wild (as far as I'm aware), even though the two species must come into contact in some places, further supports this conclusion.

on. By the fifth generation, the resulting Savannah has only three percent serval heritage. Though still handsome, these cats would not be mistaken for servals—they're browner and have a longer tail and a less pinched face.

And they're also generally smaller.* Although F1s are very large, and F2s still pretty large, later-generation Savannahs are usually more in the weight range of large cats, like my Winston. That's not to say that late-generation Savannahs look like any old domestic cat; quite the contrary, they have a distinctive spotted coat and their legs are still substantially longer than typical domestic cats.

Speaking of those legs, Savannahs are the tallest cat breed, the yin to the Munchkin's yang. From the belly up, Savannahs, just like Munchkins, appear to be normal, somewhat slender cats. But down below, Savannahs are all legs, much longer than those of the standard domestic cat. In fact, a Munchkin could easily walk underneath a large Savannah without hitting her head.

One of my favorite YouTube videos is a nine-second clip posted in 2015. The film begins with a large spotted cat, Zoey, sitting on her haunches, looking up at a light bulb on the ceiling directly above her. Suddenly, she jumps straight up, snagging the short cord and turning the light off. Stretched completely out, we see Zoey's impressively long legs. The room appears to have a rather low ceiling, but it's still an impressive six-foot-plus vertical leap. Another online video shows a

---

*Keep in mind that hybrid populations of all sorts, because of their diverse genetic heritage, tend to be more variable than non-hybrid populations or breeds. As a result, some early-generation Savannahs are small and some late-generation ones quite large. Eventually, when Savannahs have only been mated with each other for several generations, the variability in the gene pool decreases and offspring become more consistent in appearance.

different Savannah springing a yardstick-calibrated eight feet high to grab a cat toy.

Later-generation Savannahs are also tamer. There's some dispute about the suitability of F1s as house pets. Pro-Savannah websites portray them as loving, albeit rambunctious, cats, whereas detractors portray them as bloodthirsty wild animals. Certainly, at more than twenty pounds, a male F1 can be a handful. As one cat expert told me, "Those large cats can be very hard to manage as adults as they are often very confident and, thus, willing to argue with you, and you're going to lose."

Servals are among the friendlier wild cat species—there's a reason zoos use them as ambassador animals to visit school classes and other groups. A lot depends on how a particular cat is raised, of course, but my guess is that a properly socialized F1 Savannah can make a reasonably good pet if treated appropriately (they apparently need a lot of attention and, if left alone, can entertain themselves in destructive ways, just like some dogs). Still, my cat expert friend's caution should be kept in mind.

Regardless, later-generation Savannahs undoubtedly are calmer, both because they have less serval DNA and because breeders intentionally select for cats with good temperaments.

SAVANNAHS WEREN'T THE first breed created by mating a domestic cat with a wild species. That honor goes to a beautiful cat called the Bengal, a breed that got its start by accident.

Asian leopard cats certainly rival servals in good looks, with a sumptuous spotted coat, winsome big eyes, stripes running down the forehead, a little pink nose, and endearing rounded ears. However, servals were chosen as a source of a new breed not just for their good looks, but also for their friendliness and tameability. And there the comparison goes awry. Recall that zookeepers rated leopard cats as the most unfriendly of their feline charges—"a foul-tempered little beast with a gorgeous spotted coat," according to *The New Yorker*.

Still, they are beautiful, and years ago, you could buy them in a pet store despite their sour disposition. And that's what Jean Mill did. She wasn't intending to create a new breed, but when she put a black tomcat in with her female leopard cat, the two hit it off, and before long the first Bengals were born.

A Bengal.

Like Savannahs, F1 male Bengals are infertile. Backcrossing with domestic cats restores fertility around the F5 generation, or sometimes a little earlier. But unlike Savannahs, the signature feature of Bengals—dark spots on a resplendent, warm coat—only gets better with careful breeding.

Indeed, Bengal enthusiasts have done a brilliant job of breeding for enhanced spotting, building large rosettes comprised of a dark ring of connected spots enclosing an orange-brown interior. Rosettes are common in many cat species—leopards, jaguars, and margays, for example—but the Bengal was the first domestic cat to have them. How rosettes first arose in Bengals is uncertain, but one speculation is that they were derived from a mating with a blotched tabby American Shorthair, the idea being that the swirl of the blotched tabby crossed with the spotting of the Bengal somehow ended up producing a rosette.

Careful selection has also enhanced the clarity of the spots, the extent to which the spot is sharply defined against the background, as opposed to a more indistinct, gradual fade from spot to background.

In addition, many Bengals have an iridescent sheen. This "glitter trait" was first discovered by Jean Mill in a cat of unknown origin living in the rhinoceros enclosure at the New Delhi zoo. Mill finagled to acquire the cat, had her shipped back to California, and bred the trait into her Bengals (thanks to subsequent matings, Bengals have now passed their shimmer—and also rosettes—to other breeds).

At the same time, breeders were selecting for friendly cats. By most accounts, they have succeeded admirably. Early-generation Bengals often retain the ancestral leopard cat ill manners, but by later generations, the cats are affectionate and loving companions, though most definitely not lap cats.*

The result of all this breeding is an extraordinary cat that has only gotten more and more beautiful through years of breeding. As a result, the Bengal has become one of the most popular breeds today, with more than two thousand breeders existing worldwide.

But all is not well in Bengal World. Some have higher aspirations

---

*Not everyone, however, has gotten the memo. One European Burmese breeder told me she would not sell a cat to someone who owned a Bengal because the latter are wild animals that would tear her babies limb from limb.

for the Bengal, wanting to make it more than a domestic cat with a gorgeous paint job.

Jean Mill deservedly gets credits for creating the Bengal. But the man to whom she passed the baton, who has become the leader of the Bengal breeder community, is Anthony Hutcherson. And rightly so, because Hutcherson has been entranced with all things Bengal since he was a teenager.

His fascination began during a visit to his elementary school library at age eleven. Already enamored with cats, he stumbled across an old book on pet ocelots. Keeping ocelots and even larger cats as pets, including cheetahs and leopards, really was a thing in the middle of the twentieth century. But by the time Hutcherson got hooked on the idea, doing so was a lot harder and more expensive. Even as he came to appreciate that ocelots belong in the jungle rather than in the home, the idea of a spotted pet cat stuck with him. And just a few years later, he was breeding his own Bengals. His goal: if he couldn't own a wild ocelot, he'd create a domestic cat just as wild- and exotic-looking.

One more thing you need to know about Hutcherson: he's incredibly engaging and personable. He's got a great smile. And he's a great showman.

Put those traits together, add the allure of a beautiful wild-cat-looking pet, and it's no surprise that he's become a media darling. Hutcherson has appeared in *The Washington Post* and *Time* magazine. He's been featured on CBS News and *The Martha Stewart Show.* When the Westminster dog show invited cats to join their 2017 extravaganza, he was there with his cats and stole the show. When *The New Yorker* organized a Dogs vs. Cats debate, he was selected for Team Cat; a photo of him, a Bengal stretched along each arm, led off the ensuing article.

And that's not to say that he doesn't have the bona fides. He's been the chair of the Bengal cat breed section of the International Cat Association since 2009 as well as serving as president of The International Bengal Cat Society. His cat, Prestige, was TICA's top Bengal in the world in 2016; Prestige's mom, Abiding Ovation, was top kitten a year earlier.

So Hutcherson's a big wheel in the Bengal world. What's his beef? From the outset, the goal with the Bengal was to create a cat evocative of the wild. But, he argues, there's more to a jungle cat than her coat.

The TICA Bengal breed standard specifies that "the goal of the Bengal breeding program is to create a domestic cat which has physical features distinctive to the small forest-dwelling wildcats, and with the loving, dependable temperament of the domestic cat." By "forest-dwelling wildcat," they mean the Asian leopard cat or the ocelot. And Hutcherson, though he loves Bengals as they are, sees that they have a ways to go.

In comparison to other domestic cats, "ideally a bengal's ears should be more rounded than pointed and the eyes should be leaning more toward larger than small; the tail should be thicker along its length and shorter than other domestic cats and the head should be a little bit longer than it is wide," he explained. Moreover, to make the cats more ocelot-like, "I would really love to have horizontally flowing rosettes that are either connected to one another or repeating in a horizontally flowing pattern, in a manner not currently found in any domestic cats." So stay tuned: stunning as the Bengal is, if Hutcherson has his way, the breed will become even more gorgeously exotic.

There's a second issue that concerns Hutcherson. Some breeders keep clamoring to import more Asian leopard cats to add to the breed. But there's no real reason to do that. Late-generation Bengals aren't lacking any traits; if anything, they're better than early-generation hybrids. And thanks to all the outcrossing to different breeds, as well as the many Asian leopard cats that were used in the past, the Bengal already has plenty of genetic variability.

And there are downsides to bringing in more leopard cats, one that concerns Hutcherson passionately. The Bengal is now considered a domestic cat breed. Reviving concerns about their wildness would damage the breed's reputation. Plus, we don't want to jeopardize the Asian leopard cat's existence by promoting international trade in them.

Despite these concerns, some breeders want to import more leopard cats for one simple reason: money. For some, there's a cachet to saying you have a cat one generation removed from the jungle, their unfriendly disposition notwithstanding. Hutcherson would like to see that stamped out. Leave well enough alone—the Bengal is already a fabulous breed of domestic cat. Let's keep Asian leopard cats where they belong, in the jungle.

Although Savannahs and Bengals are the marquee hybrid cats, other breeds have been created by crossing domestic cats with jungle cats (producing the Chausie), Geoffroy's cats (the Safari), and caracals (the Caracat). None of these breeds is common. Rumors abound of attempted cross-matings with many other species.

Many major cat fancy organizations, in agreement with Hutcherson's position, have enacted bans on the recognition of further breeds developed from wild species out of concern over the impact of wild cat populations, many of which are already threatened from other causes. Nonetheless, that just means that those breeds can't be officially registered with those organizations; people can still do what they want, and probably will.

. . .

BREEDERS ARE PRETTY RESOURCEFUL. They capitalize on newly discovered mutations, cross breeds to combine their distinct traits, and even borrow from wild cat species, all in the service of taking advantage of variation that occurs in some cat, somewhere.

Sometimes, though, breeders don't start with a particular trait and build a breed around it. Rather, they begin with a portrait in their mind's eye, a cat they would like to see exist. To make that vision a reality, they assemble the breed bit by bit, selecting from a multitude of cats from all walks of feline life to bring together the desired traits.

From a young age, Karen Sausman was fascinated by animals. In college, she stumbled into a part-time job as a keeper at the Lincoln Park Zoo in Chicago. From there, one thing led to another, and she ended up being the founding director of the Living Desert Zoo and Gardens in Palm Springs, California. During her thirty-nine-year tenure, the Living Desert grew from a local tycoon's crazy notion to an eighty-acre institution with half a million visitors every year. During this time, Sausman served the zoo world in many ways, developing innovative practices and receiving many honors, including the R. Marlin Perkins Award, the highest recognition for professional excellence in the zoo field.

You'd think that building a zoo from scratch would leave little time for other pursuits, but on the side, Sausman was also an avid animal breeder. Over the years, she kept and bred Andalusian horses, as well as several breeds of dogs. She even dabbled in llamas, producing a national champion in only five years.

A second hobby—drawing and drybrush watercolor portraiture—was surprisingly related to the first. The commonality was a creative sense, the ability to envision what an animal might become, either on canvas or in the flesh. A self-taught artist, she developed a knack for examining her detailed compositions and figuring out what was needed to get the animal just right. "If you're trying to draw an animal true to life, you eventually develop a way of looking at things so you can really see it." And just as you can look at an animal and see how to portray it on paper, she explained, you can also look at a group of animals and see which features need to be molded or tweaked to get a breed just where you want it.

Breeding dogs is her true love, and she became renowned in the dog fancy for the new twist she created in breeds like the Fox Terrier. But cats offered an opportunity not available in dogs: the ability to create a brand-new breed by crossing different breeds (mixed-breed "designer" dogs like the Saint Berdoodle are not eligible to become breeds recognized by the major dog organizations*).

When the Savannah was first developed, Sausman was intrigued. She'd always liked the servals at the Living Desert and was a big fan of leggy dogs as well, having raised Sighthounds and Scottish Deerhounds. No wonder she was smitten! But Sausman thought that as a serval stand-in, the Savannah was too big to be a good household pet: a normal-size cat with the long-legged, big-eared look would be better. The solution was obvious: she'd create her own housecat-size version from scratch, without using any cats with serval DNA.

---

*I don't know the reason for this rule, but the internet is full of conjectures pro and con on the merits of such crossbreeding. Perhaps the rationale is akin to the International Cat Association's new rule about not allowing the use of a mutation to characterize multiple breeds. With so many dog breeds already in existence (two hundred to nearly four hundred, depending on which dog organization you ask), the number of combinations possible by crossing breeds would be enormous. Regardless, there is no genetic reason why crossing breeds shouldn't be allowed; if anything, mixed-breed dogs are likely to be healthier.

But how? There was no breed that was almost a serval and just needed the infusion of a single trait from another breed. Rather, she'd need to find the traits wherever she could and mix them together. Add a dash of long legs here, a pinch of spots there. Throw in some big ears. Large, round eyes would be good too. Hopefully, with every generation the resulting cats would get closer to the goal.

She started with Bengals, which she'd already been breeding, to get the spots. Orientals had long legs and big ears, but the legs weren't long enough and were too delicate, and the ears were on the side of the head. So instead of choosing champion Orientals to breed with her Bengals, she'd look for ones that were imperfect (in the cat fancy sense), with characteristics that didn't match the Oriental breed standard, but were just what she needed. Same with the Bengal: she'd choose ones that were most like the vision in her head.

In addition, she always kept her eyes out for other cats, pedigreed or not, that had the traits she was looking for. One was a shelter cat that had perfect long legs and large eyes. Another was a stray imported from India that had nice spots and long legs. Several additional Bengals and Orientals were thrown in over the years as well.

Once she had the ingredients, it was just a matter of selecting, generation after generation, for the offspring that were moving in the right direction, getting the cats to look more and more like the ideal cat she envisioned. "You're creating a living piece of art," she says, "painting with genetics."

Twenty years later, the result—christened the Serengeti—is beautiful: long legs; large, upright ears; a powerful, yet slender body wrapped in a warm yellow-brown coat blanketed in black spots.*

---

*Serengetis also come in an all-black model with "ghost spots" faintly visible in the right light, evocative of melanistic servals.

A Serengeti.

You're not going to mistake a Serengeti for a serval—that was never the intent. Rather, Sausman has created a bewitching new cat that pays homage to the classic African version. And as an added bonus, they're real sweeties, playful and affectionate.

So we have two serval-inspired new breeds of cats. If you're a serval aficionado, which one do you choose?

In looks, temperament, and size, late-generation Savannahs and Serengetis are very similar. The major difference is that Serengetis have larger, rounder eyes, a slightly longer head, and less robust bone structure.

A bigger difference—and frankly, the appeal to many—is the more servaline look and large size of F1 and F2 Savannahs.* The tallest

*For these reasons, another trait of Savannahs that changes from one generation to the next is the price tag: F1s can go for five figures, whereas F5s are mere pocket change at two grand, give or take a few Benjamins.

domestic cat ever measured was Arcturus Aldebaran Powers, an F2 Savannah measuring nineteen inches at the shoulder and weighing thirty pounds. Get out a ruler and see for yourself how tall that is! It's more than twice the height of the average domestic cat. These are the cats that can jump higher than a person's head in a single bound.

Remember, though, that these large, early-generation Savannahs are not a self-sustaining population. You can't mate two of them to get large offspring, because the males are sterile. As a result, the production of these cats requires continual mating between servals and domestic cats. In turn, that requires a steady supply of servals, either from the wild or from serval breeders. Although servals are not currently an endangered species, many still consider this a bad idea. Serengetis and late-generation Savannahs, of course, are completely fertile and don't present this problem.

A FEW YEARS BEFORE SAUSMAN embarked on her project, another Southern Californian hatched a similar plan to create a miniature version of nothing less than the largest cat in the world, the tiger.

That Judy Sugden had such a grand vision is no surprise. Like Sausman, she's artistic, though in a different way. An architect by training, she was imaginative enough to envision new creations and organized enough to figure out how to make them happen. Sugden also had a predisposition that Sausman didn't have. Jean Mills, the woman who created the Bengal, was her mother. As a result, Sugden had grown up sculpting new cats, and if her mom had made a little leopard, wasn't a tiny tiger the logical next step? Sugden's goal was simple. Combine the essence of the tiger—big-boned and powerful, orange with black stripes—with the character of the domestic cat—friendly.

Creating a tiger-striped domestic cat was more of a challenge than you might expect. Orange cats exist, and mackerel tabbies have vertical black stripes. You'd think all you'd need to do is mate the two to combine the traits.

But there's a catch. The allele that produces orange affects all the markings on a cat, not just the background color. Orange mackerel tabbies exist, but their stripes are not black, just a darker shade of orange.* Garfield fans take note—your idol is a biological impossibility.

And there's a second difficulty. Tabby stripes have a different arrangement than tiger stripes. Tabby striations are very linear, like bars in a jail cell. By contrast, the tiger pattern is more of an interbraiding of stripes originating from the midline of the back with others coming up from the belly. Moreover, some of these stripes branch, and some of the branches then come back together, forming enclosed spaces.† Tony the Tiger fans take note—your favorite cereal spokesman has domestic cat, rather than tiger, striping.

Sugden's playbook for creating the Toyger—as the new breed was inelegantly named—was very much the same as Sausman's. She started with a Bengal with three important traits—a tan coat, thick bones, and a loving disposition—and bred him with a stray mackerel tabby with prominent stripes. As the project proceeded, cats with other tiger-like

---

*The different colors of stripes and background is a result of the intensity of pigment deposition, rather than being produced by different pigments. Typical mackerel tabbies have black stripes on a gray background. Both the black and gray color are produced by the same pigment in the hair, eumelanin. The different colors are a result of different amounts of eumelanin deposited on hair shafts. Orange cats produce phaeomelanin rather than eumelanin; differential deposition produces darker orange stripes on an orange background.

†Why domestic cat and tiger stripes differ is unknown. My guess is that it is just a fluke of history; different mutations just happened to occur in ancestors of the two types of cats, producing different types of stripes. Of course, it's possible that there's some adaptive significance to the difference, tiger stripes being more effective for camouflage in tiger habitat, domestic cat (well, really wildcat) stripes better where they occur.

features were carefully added to the mix. The inclusion many years later of a Bengal with rosettes was particularly important because Sugden figured—correctly, it turns out—that the genes for rosettes might combine with those for mackerel tabby markings to produce elongated rosettes resembling branching tiger stripes (Bengals with rosettes didn't exist when the project started). She also added cats with slightly warmer, "orangier" tones, eventually getting the color she wanted without using the allele that typically produces orange color in housecats. Generation after generation, she chose the offspring that had the best features and kept on breeding them, getting closer, "squinch by squinch," to the look for which she was aiming.

A Toyger.

Sugden and Sausman's approach shared one more similarity. From the outset, both selected not only on anatomical and color pattern traits, but also on temperament: "My philosophy was that I was not just creating lovely cat sculptures to look at—but living creatures that could become part of a human household," Sausman said. Both went out of their way to choose friendly, agreeable cats; in fact, that was the top priority. The reason was simple. In selective breeding, you produce a lot of kittens and then choose the few best individuals as breeders for the next generation. That leaves a lot of other kittens that don't make the cut. What's to be done with the extras?

Their strategy was straightforward: if they developed cats that were extremely friendly, they'd have no problem finding homes for loving, B-grade Toygers or Serengetis.* The result of this selection, at least if you believe fawning cat websites, is that the two are among the most affectionate of cat breeds.

It's now been several decades since Sugden and Sausman set out to create new types of cats, long enough to evaluate how successful they've been. On the positive side, both women have created distinctive breeds of cats. The Toyger, with its bold orange and black markings, is readily identifiable as a tiger wannabe, and a darling one at that; Toygers look like little cats in tiger pajamas.

The Serengeti does justice to its name, an elegant, long-legged, spotted cat with great big ears on top of its head, seemingly as much at home on the plains of Africa as in your living room.

But both women agree that their breeds are works in progress. Other than the stripes, Toygers do not evoke "the essence of tigerness" Sugden's been aiming for—clothes may make the man, but stripes do not make the tiger. Many of the tiger traits she's targeted—big, round head, small ears, circular striping around the side of the head—remain to be attained.†

Similarly, Serengetis have a way to go as well. Sausman highlights the clarity of the coat—the extent to which the spots jump out dis-

*Things have not always been so genteel in the animal-breeding world. Darwin remarked in his book on domestication, "Lord Rivers, when asked how he succeeded in always having first-rate greyhounds, answered, 'I breed many, and hang many.'" Some unscrupulous breeders today are suspected of behaving similarly.

†Though days before I finalized my manuscript, Sugden reported the toyger community had recently achieved a great breakthrough in producing cats with a long face, straight nose, and small eyes, all more typical of tigers than domestic cats.

tinctly from the background—as an ongoing challenge. Also, although her goal has always been normal-size cats, heavier boning is needed because some Serengetis look too refined.

Undoubtedly, one factor that has kept progress slow is the scale of the operation. Sausman, for example, generally had only six to ten breeding adults at any one time. Evolution by natural selection requires variation, and the larger the population, the more variation there will be from which to choose. There's a reason that fruit-fly researchers use populations in the thousands for their evolution experiments. I'm amazed at how quickly cat breeders can transform their breeds while working with such a small number of cats.* And I suspect that, given enough time, Toyger and Serengeti breeders can get the breeds to where they want them to be.

NOT EVERYONE IS ENAMORED with the creation of new cat breeds, and for good reason. Consider squittens, aka twisty cats. These are cats born with extremely small or absent forearm bones. As a result of this deformity, the cats spend most of their time sitting upright on their hindlegs, like a squirrel; they often hop around like a kangaroo because their ability to walk on all fours is impaired. Kittens have trouble kneading their mothers' belly to get milk; adults have trouble burying their feces and carrying out other normal activities.

Twisty cats arise periodically by random mutation. Many die or are euthanized at a young age; in some cases, good-hearted souls raise these cats and give them as good a life as possible. After neutering them, of course.

*Of course, the fly's much shorter generation time is also responsible for the much quicker rate of evolution in lab fruit-fly experiments than in cat breeding.

In the late 1990s, the world was outraged when word got out that people in Texas had intentionally bred these animals because they thought twisty cats were cute and wanted to make more of them. How could anyone be so heartless as to intentionally produce cats cursed to live life with such a deformity? Fortunately, the plan to create a breed of twisty cats was abandoned.

But are twisty cats all that different from Scottish Folds?

In 1961, a female cat with her ears lying flat against her forehead was found on a farm in Scotland. Susie subsequently had a litter that at first seemed normal, but after several weeks, the ears of two of her kittens tipped forward as well. The resulting owl-like appearance of these cats was adorbs, and the Scottish Fold was born. Subsequent breeding established that the trait is due to a dominant mutation that adversely affects cartilage and bone development; as a result, the cartilage is not rigid enough to hold the ears erect.

If the only effect of this mutation were to cause folded ears, no one would care. The cats are cute and quite sweet. But in fact, the mutation affects cartilage and bone throughout the body. Cats homozygous for the allele are severely affected, with misshapen toes and small feet, thick and inflexible tails due to fusion of bones, and severe progressive arthritis. Although many of the health problems do not occur in heterozygotes, the evidence suggests that all have arthritis, some more seriously than others.

Breeders mate heterozygotes to cats lacking folded ears, thus producing litters composed half of folded-ear heterozygotes and half of straight-eared cats without the allele. Although this approach prevents the production of homozygotes doomed to live a miserable life, the fact is that many heterozygotes develop symptoms, sometimes badly compromising their quality of life. As a result, the major European cat organizations do not recognize the Scottish Fold as an accepted breed. In addition, breeding these cats has been outlawed in some parts of Europe.

A similar situation exists with the Manx. Originating more than two centuries ago on Great Britain's Isle of Man, Manx either lack a tail or have a shortened one (Manx are categorized by tail length, from tailless rumpies, through rumpy risers and stumpies, to relatively long-tailed longies). Unfortunately, the dominant gene responsible for this tail abbreviation also causes serious deformities, such as incomplete formation of the spinal cord, fusion of the spine, and problems with the colon. These afflictions are so severe that many have suggested that if the Manx breed didn't exist and someone tried to create it today from a newly arisen mutation, the public would be outraged and cat organizations would refuse to recognize it.

The traits that define the Scottish Fold and the Manx are the cause of serious health issues. The situation is different for the Persian, which has existed for more than a century and for much of that time was not burdened with an unhealthy physique. Persians weren't always nasally deficient. In the old days, they had perky, smallish noses and didn't have any particular health issues. But since then breeders have pushed to eliminate their noses and today's flat-faced Persians are plagued with serious problems.

The reconfiguration of the face and skull leads to a variety of dental, breathing, and respiratory issues, as well as preventing normal operation of the tear ducts. German and Swiss researchers used MRIs and CAT scans to examine the skulls of ninety-two Persians and determined that "in the modern peke-face type, the increased reduction of the nose is associated with . . . the occurrence of severe skull and brain abnormalities." Indeed, Persians are widely regarded as not the sharpest knives in the drawer and have been observed running into objects and falling off windowsills; the researchers suggested that brain reorganization required by the skull shape of Persians was responsible for their dimwittedness. The scientists concluded their paper: "Breeders and cat fanciers must face the fact that desired . . . traits in these cats would be considered a severe developmental abnormality in humans."

Research in the UK provided supporting evidence. In a study of nearly three hundred thousand cats that had received veterinary care in the United Kingdom, the three thousand Persians had a substantially higher incidence of a number of disorders than other breeds. Many of these health issues were plausibly connected to their skull shape: the link to eye problems is clear, and the altered shape of their jaws and resulting dental issues may be the cause of Persians' skin and coat ailments because the cats are unable to effectively groom their trademark long hair (on the other hand, the average life span of Persians was no different than that of other pet cats, pedigreed or otherwise).

For these reasons, several European veterinarians have suggested the Germany's animal welfare law should be interpreted as prohibiting the breed of cats in which the top of the nose is above the lower eyelids. More generally, veterinarians, animal welfare advocates, and scientists around the world have called for reform and regulation of the breeding practices that have produced the modern-day Persian.

In all these cases, breeders claim that by only breeding healthy individuals, they can select for the positive traits they favor—the folded ears, the flat faces, the absent tails—while excluding the undesirable traits and thus not compromising the cat's health. But as the traits they prefer and the ones they'd like to eliminate are strongly linked, the result of the same genes, there are real doubts whether they can be dissociated. There is no gene specifically for cartilage in the ears, for example; the genes affect cartilage everywhere. As a result, it is probably impossible to breed individual cats with folded ears but otherwise perfect cartilage.

On the other hand, not every unusual trait creates serious problems. Consider hairless cats. Some have proposed they should be banned because they'll get cold outside. If, in fact, Sphynx were regularly being kept outdoors in Minnesota in the winter, that might be a problem, but it seems likely that few people who go to the trouble and expense of getting such a cat would be so stupid. Extreme traits of other breeds may seem problematic—think about the tiny legs of Munchkins or the long, narrow face of Siamese—but evidence of health problems is lacking.

There is a second ethical issue regarding pedigreed cats. With the great number of cats in shelters and foster care, how is it morally defensible to breed more cats? Instead of buying Persians and Siamese, everyone should adopt a rescue cat! My sister has obtained all her cats from shelters, looking especially for black cats, which are less readily adopted. Countless others take the same approach.*

It's true that there are many available cats waiting to find their forever home (though, thankfully, not as many as there used to be due to the hard work of animal advocates). On the other hand, there are good reasons why some may want a pedigreed cat. In particular, some people would like to live with a cat that behaves in a particular way. And as we discussed in chapter twelve, some breeds are characterized by specific behavioral tendencies.

A few years ago, we decided to get a cat for my father. It was important that we find one that was affectionate, friendly, and not too boisterous, a good companion for an eighty-five-year-old. Melissa did some

*Of course, plenty of people with pedigreed cats also adopt shelter cats. And the cat fancy organizations themselves contribute generously to cat-support organizations.

research and came up with a short list of breeds that were particularly known for these traits. By good fortune, we soon learned that the European Burmese Rescue Network had an EB that needed a new home; in short order, Aten* was living with my parents.

Aten is everything we hoped for. Remarkably loving from the moment he arrived, Aten quickly won the hearts of the entire family (including my very skeptical mother, hardened by years of furniture scratching by poorly trained previous pusses). Indeed, it was because of our experience with Aten that Melissa and I obtained Nelson a year later.

Others have different criteria for the type of cat they'd like. For those looking for a docile cat—sedate, not too active or vocal, definitely not a furniture scratcher, an all-around good lap cat—Persians are often the choice; their zen master–like tranquility explains the appeal of this breed to many. Ragdolls, which have surpassed Persians in popularity in recent years, have the same qualities without the Persians' smooshed-in nose and attendant health issues.

On the other hand, some people want an active cat, one full of energy and always looking to play. For them, an Abyssinian or a Bengal might be the way to go. And if for some reason you really want a talkative cat, there's no doubt that a Siamese (or related breed) would be your choice.

Of course, there are cats in shelters and other adoption sites that have all these qualities. But the thing about breeds is that they're generally consistent in their characteristics. Get a Bengal or a Persian and you've got a good idea what she will be like (naturally, there are exceptions). Non-pedigreed cats are much less predictable (though,

---

*Named after an Egyptian sun god. Yes, my dad had a penchant for unusual cat names.

of course, the more you acquaint yourself with the cat before adopting her, the better an idea you'll have of what you're getting). You may also be hard-pressed to find randombred cats with personalities as extreme as some pedigreed cats. Few cats are as placid as Persians, as vocal as Siamese, or as active as a Bengal.

These points notwithstanding, some are strongly opposed to purchasing pedigreed cats given all the cats needing homes. The view is part of a broader discussion about how charitably and altruistically people should act. This is a debate we're not going to solve here; it's part of a larger discussion about consumption, resources, and ethical decision-making.

There is, however, one controversial practice in creating pets that are more "convenient" to domestic life about which there should be no debate: declawing and similar procedures are mutilation and are morally indefensible. The term "declawing" may suggest a simple procedure of removing the claw from the toe, perhaps barely more than clipping your toenails. But in fact, it's much worse: declawing entails amputating the toe bone to which the claw is attached. Think about taking garden shears and cutting off the last bone on each of your fingers—that's what declawing is. The procedure can produce lasting pain and causes a wide variety of physical and behavioral problems. Instead of inflicting suffering on your feline friend, take the responsible, humane course: buy a scratching post and train your cat!

Fifteen

Catancestry.com

Havana Browns are unusual-looking cats. Brilliant green eyes contrast sharply with their rich, mahogany coat, all set on a slender, pantherine body. But what really gets your attention is the schnoz, described as a corncob or the bottom end of a light bulb. As a result, the face of a Havana Brown seems divided into two parts—a long tubular muzzle glued onto a typical wedge-shaped head. No other cat has a visage anything like it.

Havana Browns date to the early 1950s, when several Brits decided to create a breed of brown cats by mixing Siamese and black cats, perhaps adding in a sprinkle of Russian Blue. The result was a beautiful cat breed colored more warmly and richly than the sable brown of Burmese cats. As for how they ended up with their trademark snout, no one seems to know—one of the quirks of breeding lost in the fog of time.

Havana Browns have never been particularly popular. By the 1990s, there were only twelve registered breeders in the world. Inbreeding was becoming a problem. Selection to improve the breed was impossible—it was hard enough to find cats to mate that weren't closely related, much less to choose mating pairs based on desirable traits. The future genetic health of the breed seemed questionable. The breed's leaders realized they needed to consult a feline geneticist. Leslie Lyons, a researcher then at the National Cancer Institute, is who they found.

A Havana Brown.

As occupations go, cat geneticist is a pretty obscure choice. How does someone end up spending her life researching cat DNA?

According to one crazy-sounding theory called "nominative determinism," people gravitate toward occupations that agree with their names. Many dentists are named Dennis, for example. Similarly, people named Butcher are more likely to be meat vendors than to take on other occupations, while those named Miner, Baker, Barber, and Farmer are each more likely to work at their corresponding trades.

So it seems fair to ask whether Lyons was destined for a career with felines. She admits to always having liked cats, but says that her career path was more a fluke than nominatively predestined. Initially interested in going to medical or veterinary school, she got hooked on genetics in college. She recalls watching Jerry Lewis telethons and thinking, "I want to solve muscular dystrophy." Off to grad school she went to get a PhD in human genetics, studying the genes causing several human diseases, including a type of colon cancer.

As her grad school days wore on, she met researchers studying animal genetics and found the topic interesting and the field underpopulated. If the FBI had been more on the ball, they would have hired her when she applied for their new animal forensic genetics position—CSI Animal Lab—but criminology's loss turned into feline research's gain.

Lyons eventually joined the laboratory of Stephen O'Brien, the same lab where Carlos Driscoll would get his start a few years later. O'Brien's group studied many species, and Lyons didn't know what she'd be studying. But when she arrived, O'Brien told her: cats, an assignment that set her down the path to becoming the world's leading authority on feline genetics.

An initial project involved the Asian leopard cat. At the time, researchers thought that viruses might play an important role in cancer. The Asian leopard cat is immune to feline leukemia virus; they hoped that by understanding this immunity, they could learn something applicable to human cancers (this is why O'Brien's National Cancer Institute lab started working on cats).

The lab was working with the National Zoo in Washington, D.C., trying to mate leopard cats with domestic cats to produce study subjects with genes from both species. One day, while thumbing through the ads at the back of *Cat Fancy* magazine (which Lyons had taken to reading to get up to speed on all things feline), she discovered that people were selling hybrids from crosses between Asian leopard cats and domestic cats. She had an epiphany: why go to all the trouble of breeding them yourself if the hybrids were already being produced elsewhere? She schmoozed the breeders, who turned out to be very willing to let her get genetic samples from their cats.

The biomedical work didn't cure cancer—we now know that viruses are responsible for very few types of cancer—but the lab's genetic studies of the hybrids represent the dawn of modern-day feline genetics, providing the first "map" of the feline genome and revealing that cat and human genomes are organized in a very similar way.

In addition to the scientific results of the project, the connections Lyons made served as her entrée into the world of the cat fancy. Just a few years later, when Havana Brown breeders became concerned about genetic issues, they naturally turned to her for help.

By analyzing genetic samples provided by the breeders, she was able to show that Havana Browns had substantially less genetic variation than typical, randombred cat populations. Inbreeding problems—such as small litter sizes, immune-system deficiencies, or high incidence of inherited diseases—were a distinct possibility. Lyons advised bringing in much-needed genetic variation by establishing crosses with cats from other breeds. Her recommendations were accepted, and Havana Browns—still with the distinctive schnoz—are now more popular than they've ever been (though that's a pretty low bar).

Advising cat breeders, however, was not the highest priority of Lyons's job. Rather, she became increasingly focused on trying to identify the genes responsible for feline diseases.

IN THE OLD DAYS, the way we learned about the genetic control of a trait—anatomical, behavioral, or otherwise—was by breeding individuals and observing their offspring. Wavy-haired rex cats provide a great example.

In 1950, an unusual cat was born in Cornwall, England. Kallibunker, as he was christened, had a variety of unusual traits including a lithe body with long, delicate legs, a wispy tail, and an elongated, narrow head. But what really stood out was his hair, arranged in marcel waves like those produced by a hot curling iron and fashionable in the 1920s. Kallibunker was the founding member of a new breed, the Cornish Rex.

The name "rex" may suggest some regal aspect of their appearance, perhaps the unusual Roman nose that characterizes Cornish Rexes today. In fact, the name has nothing to do with royalty. Rather, the fur of these cats is very similar to that of rabbits said to have "rexed" fur. How the rabbits got that name is unclear. Regardless, wavy-haired cats have borrowed the moniker and ever since have been called rex cats.

When Kallibunker was mated with his mother,* both curly-haired and straight-haired kittens were produced, but when he was mated with other cats, there wasn't a marcel wave to be seen among the offspring. Further matings confirmed that the rex hair trait must be recessive, requiring two copies of the rex allele to produce marcel-waved cats. Kallibunker's straight-haired mom must have been a heterozygote, a carrier for the rex hair trait.

Ten years later, a cat was born in nearby Devon with a somewhat similar wavy coat. Kirlee was considered to be a rex cat as well and assumed to share the same rex allele as Kallibunker and his clan.

Because rex hair is a recessive trait, mating two rex-haired cats must produce nothing but rex-haired offspring: the kittens will get one rex hair allele from their mother and the other from their father.

To everyone's surprise, that's not what happened when Kirlee was mated with some of Kallibunker's wavy-haired descendants. Quite the contrary, all of the offspring—all of them!—had normal, straight

*Ick! But it was possible that Kallibunker's mom had the allele for wavy hair as well, hence the oedipal arrangement (remember that for recessive traits, offspring must inherit the allele from both parents to exhibit the trait).

hair. Imagine the consternation of the breeders! At first, this seemed like a fluke, but when the same result occurred consistently, the meaning was clear: the wavy hair of rex cats from Cornwall and Devon was the result of different genes that produced similar effects. Offspring of crosses between the two ended up heterozygous for both genes and thus had straight hair (rex hair is a recessive trait for both genes). And thus, the Cornish and Devon Rexes were recognized as distinct breeds.

The opposite scenario played out with another wavy-haired breed, the German Rex. Again, the rex hair trait was recessive. In this case, however, when Cornish and German Rexes were crossed, all their offspring were wavy-haired: the two breeds, different in other characteristics, shared the same rex allele.

For many years, this sort of pedigree analysis, very similar to Gregor Mendel's famous experiments with peas, was a standard approach to studying the genetic basis of traits. In recent years, the genome science revolution has allowed geneticists to go a lot further. Instead of just asking whether two breeds share the same allele, geneticists can now decode the DNA makeup of the responsible genes.

In theory, discovering the genetic difference that causes, say, the Munchkin to have short legs should be simple. Just sequence the genome of a Munchkin and some normal-legged cats, look up in the cat gene database which gene is responsible for leg length in cats, then compare the DNA bases in that gene between the Munchkin genome and that of the normals.

The problem is that when we say the genome of something has been sequenced, what we mean is that we have determined the arrangement of all the DNA bases—two billion-plus in cats. So we have a long alphabet string of As, Cs, Ts, and Gs (standing for the four kinds of DNA bases: adenine, cytosine, thymine, and guanine). Looking at this sequence of letters, we usually can identify where one gene stops and the next one starts. But even though we can locate where the genes are in the genome, we don't know what most of the genes do—the genome doesn't come with an index. As a result, if we're looking for the gene or genes that produce some particular trait, we usually don't know where in the genome to find them.

Nonetheless, geneticists are a clever lot. Even though it's a bit like looking for a needle in a haystack, they've developed ways to identify the genes affecting a particular trait. To find the gene that causes short legs in Munchkins, for example, all you need to do is sequence the genomes of a bunch of Munchkins and an even larger and diverse group of other types of cats. Then compare the DNA of all the cats, scanning the entire genome to look for a gene in which the DNA of all the Munchkins is the same and different from that of all the other cats. It's a lot more complicated than that in detail (for example, there may be several genes for which Munchkins differ from all other cats, so figuring out which one is responsible for short legs may be difficult), but in essence, that's how you can find a gene responsible for differences among individuals.

In the Lyons' Den, as her lab is known,* they did just that, finding the gene responsible for Munchkins' diminution. They also located the genes for wavy hair, confirming that the Cornish and German Rex share the same allele of a gene different from the one responsible for the Devon Rex's curl. In a surprise, they also found that another wavy cat, the Selkirk Rex, owes its ringlets to the same gene as the Devon Rex, though they have different alleles. Breeders could have determined that themselves if they'd ever crossed the two breeds, but the Selkirk is a rare breed and apparently no one had ever tried to do so.

---

*Initially at the University of California, Davis. When the lab moved to the University of Missouri, some wags suggested it should be rechristened the Tiger's Den given Mizzou's striped mascot. Needless to say, that suggestion went nowhere.

The same approach has been taken by Lyons and other researchers to identify the genes underlying cat diseases (a disease, after all, is just another trait of a cat or any other living creature). A major early success involved polycystic kidney disease, a malady that afflicts both cats and people. PKD causes the formation of cysts in the kidney and elsewhere, leading to renal enlargement and, ultimately, kidney failure. Soon after arriving at UC Davis, Lyons focused on PKD because of its extreme prevalence in Persian cats. At that time, thirty-eight percent of Persians—three out of every eight—had the disease. And given that Persians were then the most popular breed of cat, that made PKD the most prominent inherited disease of cats.

In the same general way just described for short-leggedness, Lyons scanned the genomes of Persians with and without PKD. One gene—*PKD1*—stood out as exceptionally correlated with the disease: all forty-eight cats in the study with the disease were heterozygous and had a particular allele of that gene, whereas all thirty-three without the disease did not have that allele. No individuals were homozygous for the disease allele, which suggests that homozygotes die as embryos.*

Once the gene was identified, Lyons's lab developed a diagnostic test for cats carrying the allele. The importance of this test can't be overstated. One benefit is that cats can be tested to determine whether they are likely to develop the disease, allowing preventive measures—such as special diets or more frequent wellness tests—to be put into place. Even more importantly, breeders can determine whether their cats possess the allele even before the disease develops and, if so, can

*A situation that is true for many genetic diseases, and sometimes occurs as well for alleles that are not harmful as heterozygotes, such as the allele that causes short limbs in Munchkins.

stop them from breeding. As a result of these now widespread screening measures, the prevalence of the disease has plummeted to less than ten percent of Persians. Score one for genetic testing!

Lyons and many other researchers have now identified the genes causing almost one hundred cat diseases. Another forty-four alleles underlying desirable traits like long hair and white gloving on the feet have also been identified. This is a vibrant area of research, and the pace of discovery is increasing rapidly.

For humans, this is the age of personalized medicine, in which people's genetic makeup informs their medical care. Similarly, for cats, the era of felinized veterinary care is dawning as well—it's now possible to test your cat for a battery of genetic diseases. Already, the entire genome of your cat can be sequenced for about six hundred dollars, and the price is dropping steadily. With a genome in hand and our steadily increasing knowledge of the genetic basis of many diseases and conditions, it will soon be possible to take a "P4" approach to feline health care: predictive, personalized, preventive, and participatory. Lyons's 99 Lives Cat Genome Sequencing Initiative has played a significant role in developing these capabilities by sequencing the genomes of more than three hundred cats and identifying more than seventy million different genetic differences among cats.*

These studies also hold the promise of benefiting human health.

---

*You can support this initiative by paying to have your own cat's genome sequenced! About twenty percent of the 99 Lives project is financed by people paying for particular cats to be sequenced. Some are breeders trying to locate a gene of importance to their breed (currently, for example, breeders of the bobtailed Highlander are looking for the gene that affects tail length). But others are just altruists helping the initiative and tickled to have Milo's genome included, with the understanding that the more cat genomes the 99 Lives initiative has in its database, the greater the scientists' ability to identify the genes underlying feline disease.

The cat and human genomes are surprisingly similar; not only do we share many genes, but in many cases the same gene causes similar diseases in both species. For example, a similar mutation in the PKD gene in humans causes a disease very similar to PKD in cats.

Both species can benefit from this similarity. On the one hand, we can often predict what gene may be causing a disease in cats by reference to the similar gene in human. Indeed, such cross comparisons have helped identify several feline-disease-causing genes. On the other hand, information flows both ways. New dietary treatments for PKD in cats are now being investigated in humans as well. And Lyons's group has suggested that the gene they've discovered that causes short legs in Munchkins might be responsible for similar conditions in humans. Not surprisingly, Lyons strongly advocates for more genetic research on cats to help solve other human genetic mysteries.

EVEN AS LYONS'S work on the hereditary basis of diseases and traits developed, her interest in the genetics of cat breeds also took off. She needed data on genetic variability within breeds, so she became a regular fixture at cat shows, getting mouth swabs from many participants between their turns in the ring. Increasingly recognized as the go-to cat geneticist, she was able to get breeders around the world to send her samples.

Lyons decided to expand her sampling of randombred cat populations as well. As a result, even as she was collecting data from cat breeders, she was also amassing an enormous number of DNA samples from street and pet cats around the world.

Many of these samples came from colleagues and friends, but some she gathered herself. *National Geographic* filmed her team in Egypt at work in Cairo bazaars and ancient temples. Catching cats in tourist areas turned out to be easy—the cats let the team walk right up, grab them, and quickly swab the inside of their cheeks with a high-tech cotton swab. Indeed, some cats were so relaxed, she didn't even need to pick them up; all she had to do was insert the little stick into the unrestrained cat's mouth and twirl it around (think COVID-19 testing and your nose).

In other areas, though, warier felines were less cooperative. Lyons was up to the challenge, devising a hands-off strategy to get the DNA sample without touching the cat—simply skewer a bit of meat on the end of the swab stick. When the cat mouthed the meat, it left a DNA-laden saliva sample in exchange.

Other collecting sessions were equally exotic. While on a Mediterranean cruise with her cousins, she ordered an extra plate of salmon at dinner one night, then used it to attract stray cats at a market in Tunisia the next day. In Bangkok, she taught a motorcycle taxi driver to collect samples, thirty-five of which he turned over to her as he drove her to the airport for her flight home.

Lyons's surveys showed that the Havana Brown wasn't the only breed with low levels of genetic variation: Birman, Burmese, and the Sokoke Forest cat were similar, and Singapura were even more genetically deficient. For most of these breeds, the explanation for their limited variability is obvious—the breeds originated from a small number of founder individuals and haven't been augmented subsequently by outcrossing.

In humans and many other species, numerous harmful genetic conditions exist, but the alleles for these conditions are usually very uncommon in the population. However, because many cat breeds are initiated with one or a few founding individuals, whatever detrimental alleles those particular individuals carry may end up at high frequency within the breed (this is an example of a founder event discussed in chapter eleven). This possibility is magnified when foundation animals include parents or siblings, because close relatives may have the same detrimental alleles.

Indeed, as might be expected, breeds of cats, as well as dogs and other species, often have more genetic diseases than non-pedigreed populations. Breeders work hard to eliminate such alleles from their population by not breeding animals that exhibit the detrimental condition, as well as by mating their animals with other breeds or randombred individuals to introduce new genetic variation into the population. Now with genetic tests capable of identifying the carriers of these alleles, many genetic diseases are becoming much less common, as we've already seen with PKD.

Lyons's study showed that randombred populations around the world uniformly had substantially greater amounts of genetic variation than most breeds. Only a few, like the Siberian, Norwegian Forest Cat, Japanese Bobtail, and Sphynx, were comparably diverse, perhaps due to the large number or genetic heterogeneity of cats used to establish those breeds.

Lyons's large geographic sample allowed her to test the prevailing ideas about the history of cats discussed in chapter eight. If, as the standard story goes, the domestic cat arose in the vicinity of Egypt and Turkey and from there headed north to Europe, east through Asia, and south into Africa, then genetic similarity of populations should mirror this history.

But of course, common wisdom could be mistaken. Indeed, Lyons secretly harbored the hope that she would discover a second area of domestication, perhaps in China or the Indus Valley of Pakistan, both places where agriculture emerged early in human history, potentially setting the stage for feline domestication.

Even if the conventional history was correct, however, it might not be mirrored in cat genetics. For one thing, the geographic expansion of cats happened in just the last three thousand years or so, an evolutionary blink of an eye. Maybe that wasn't enough time for genetic differences to evolve in different areas. If not, cats worldwide might be genetically indistinguishable.

And another problem: as we all know, cats get around. Even if genetic differences did evolve in various parts of the world, the transport of cats back and forth in recent times might have homogenized these differences. Consider, for example, the cats of Cairo. If present-day cat populations were a reflection solely of the history of an area, then all of Cairo's cats should look like those painted on ancient tomb walls. But in fact, Cairo's cats have colors and patterns like those anywhere else—torties, tabby mackerels, gingers; black cats, tuxedo cats, white cats; gray cats, and calicos. How did the descendants of the pharaohs' pusses get so variable?

It's possible that the mutations for these different colors and patterns all occurred in Cairo cats over the past several millennia. But it's much more likely that the alleles arose elsewhere and cats bearing them were brought to Cairo. The fact that cat populations in many places share the same apparel suggests that human transport of cats—"gene flow" is the technical term—may have acted to homogenize regional gene pools, minimizing the amount of genetic differentiation among populations.

So it's a reasonable hypothesis that few genetic differences might occur among cat populations. I do need to immediately qualify this idea, because we do know that some differences exist. In particular, the distinctiveness of eastern Asian populations—the body form and coloration of Siamese and Burmese cats, an abundance of tail-challenged cats, and a paucity of blotched tabbies—indicates that interchange of cats between that region and the rest of the world may have been limited historically. The cobby physique of European cats might also suggest limited genetic interchange with cats elsewhere, or it might just indicate that natural selection is so strong that it overrides the homogenizing effect of ongoing gene flow.

To test these ideas, Lyons (and a long list of collaborators, including Carlos Driscoll) looked at DNA from nearly two thousand randombred cats. As hypothesized, they found that cat populations the world over are pretty much the same genetically. That's not to say there isn't genetic variation among cats. Quite the contrary, there's a lot of difference genetically from one cat to the next. But very little of the variation is related to geography; most variation occurs within populations, not between them.

But the small amount of variation that does exist between populations is organized in a coherent way. If you visualize the genetic differences among cat populations as a two-dimensional diagram, right in the center are the cats of Egypt, Cyprus, Jordan, and other Mediterranean countries. On one side are the cats of Europe (including those in the Americas and Australia, which came from Europe). On the other side, and overlapping less with the center, are the cats of East and South Asia. In a third direction, closer to the Asian than the European cats, are the cats of Africa and western Asia.

In other words, the small genetic differences that do exist among populations paint the history of the cat diaspora: separate migrations to Europe, Asia, and Africa, with the eastern Asian cats particularly distinctive. Support for this scenario comes from another aspect of the data: the greater the distance separating two populations geographically, the more different they are genetically. Such an isolationby-distance pattern is consistent with a homogenizing effect of cats moving back and forth; the farther apart two places are, the less they exchange cats and their genes.*

And, finally, in the exception-that-proves-the-rule category: several locations where the cats have surprising genetic affinities to European cats—Tunisia, Kenya, South Africa, Pakistan, and Sri Lanka—are places where the British had a strong colonial influence. Presumably, the Brits imposed their cats on the locals, just as they imposed so much else.

PUT YOUR FINGER RANDOMLY on a world map and the odds are good that you'll land on a place that has a cat breed named after it. Roughly two thirds of cat breeds have a geographic label in their appellation. How boring! I much prefer the breed names Korat and Minuet to "American Wirehair." It turns out, though, that cat breeders are much more clever than I realized—many breed names mean less than they seem.

For example, I naively assumed that the Somali is from Somalia. Wrong! The Somali is a beautiful, long-haired version of the Abyssinian. Abyssinia—now called Ethiopia—borders Somalia to the south, so those creative cat fanciers picked the next country over for their newly developed breed, geographical proximity of the names signaling visual proximity of the cats. Same thing for the Javanese and Balinese, long-haired versions of the Siamese.

Havana Browns have nothing to do with Cuba (though some claim that the name stems from the similarity of their color to that of a Cu-

*The geography of human genetic variation shows very similar patterns to those seen in cats!

ban cigar). Himalayans are a type of Persian cat with Siamese points. Overthinking this, I surmised that the breed is so named because the Himalayas are halfway between Thailand and Iran (i.e., Siam and Persia). But in fact, the name is borrowed from a similarly pointed rabbit breed. How the rabbits got that name is, once again, not clear.

Finally, the Bombay does not hail from India. Rather, the woman who created the breed by crossing a black American Shorthair with a Burmese thought the resulting black cats reminded her of the legendary black panthers of India, hence the name.

In addition to these misnomers, Lyons's data from about a thousand pedigreed cats reveal other cases where genetics and reputed geography do not match. The most egregious is one of the world's most popular breeds, the Persian, fabled to be from the cold mountains of Persia. The genetics, however, tell a different story. Lyons's work shows that Persians and related breeds are genetically most similar to randombred cats and breeds from Europe (or its former colonies), rather than showing affinity to cats of southwestern Asia.

This discrepancy is easily explained. Once long-haired cats from Persia showed up in Western Europe, perhaps introduced by the Italian explorer Pietro Della Valle in the early 1600s, they became very popular and were known as "French cats."* Interbreeding with domestic cats endemic to the region led to the cobby body form of today's Persian, which is very unlike the more svelte conformation of other cats originating from southwestern Asia, like Turkish Vans and Turkish Angoras. This interbreeding was so extensive, in fact, that the infused European genes overwrote the genetic mark of Persian ancestry.

The same is true of Abyssinians. Are Abys really the descendants

*Because in the nineteenth century, many such cats were imported from France to England.

of cats brought home by British soldiers serving in Abyssinia? And if so, were the cats really from Abyssinia, or did British soldiers first transport them there from garrisons in India? That's one story, but we'll never know. Massive interbreeding with British cats during the breed's development has left them with a strong European genetic signature, obliterating any trace of their colonial past.

These and a few other examples aside, the geography of most breed names is accurate. Siamese, Burmese, Tonkinese, and Orientals do cluster together as an Asian branch of the cat family tree, genetically similar in Lyons's analysis to randombred cats from eastern Asia. Similarly, breeds reputedly from Western Europe—Russian Blues, Manx, British Shorthairs, Scottish Folds, Norwegian Forest Cats, Chartreux—cluster genetically with randombred European cats. Turkish Angoras, Turkish Vans, and Egyptian Maus are similar to cats from Turkey, Cyprus, and nearby regions.

I WAS FIRST INTRODUCED TO Leslie Lyons by a mutual friend seven years ago when I was preparing for my freshman class. In an act of genetic generosity, she said that if I could get the students to swab the inside of their cats' cheeks and send her the Q-tips, she'd run the samples in her own lab. Free of charge! That was an offer I couldn't refuse.

So we got to work. I included a sample from Aten, my father's European Burmese, and samples from the Maine Coons of an Egyptologist colleague of mine, who gave the class a cat-oriented tour of Harvard's Semitic Museum. Off went the swabs to the Lyons' Den.

A couple of weeks later, Lyons sent back the results. Aten was legit—he carried an allele only found in Burmese cats. Ditto for the Egyptologist's Maine Coons. And a surprise: one of the students' cats, Jellybean, also had a Maine Coon–specific allele. Although one look at the Beaner confirmed she was not one hundred percent Maine Coon, she obviously had a member of that breed as an ancestor.

Lyons also tested for a wide variety of genetic diseases, and the results were almost completely good news. The only blemish was that another student's cat, Bubbles, had an allele associated with blindness late in life. The trait is recessive and the test couldn't determine whether Bubbles was a homozygote or a heterozygote. Hopefully the latter.

There was even one long-term positive result of the class exercise. When Jellybean was taken in for her yearly checkup, the vet said she was overweight and instructed her owners to put her on a diet. "But wait," they responded (as I imagine the conversation), "JB is part Maine Coon, the heaviest breed of cats, so her weight is perfectly normal and no need for concern." The vet relented and the Bean's culinary regimen remained intact.*

This kind of genetic testing applied to people has become familiar to many through companies like 23andMe, AncestryDNA, and others. The tests work pretty much the same way with people as with cats: provide a saliva sample, then find out whether you have alleles for specific genetic traits and diseases. In addition, the companies provide a breakdown of what proportion of your ancestry comes from different parts of the world. 23andMe's website provides an example of someone whose ancestry was determined to be thirty-seven percent British, twenty-five percent Eastern European, twenty-two percent northwest European, twelve percent French-German, and a smattering from other regions.

---

*This is not to downplay the issue of obesity, which is a big problem with pet cats. Maine Coons, however, are much larger than your average cat, so a weight that would be alarming for most cats would be typical of Maine Coons.

This ancestral inference is possible because we humans, like cats, have a geographic signature to our genetic variation. By comparing a person's DNA to large samples of people from different regions, statistical analyses can infer what proportion of a person's ancestry comes from each region.

In recent years, several companies, as well as the UC Davis School of Veterinary Medicine, have developed comparable businesses for cats. In considering these services, I need to make one thing clear: they are based entirely on the work of Lyons and other cat geneticists. In other words, they are taking the discoveries these researchers have made—identifying the genes responsible for particular traits or diseases, establishing patterns of geographic differentiation—and commercializing them. Lyons notes wryly that these companies vary in how generous they are in giving back by supporting further research of the sort their business is based on.

Just like 23andMe, these labs test for the large number of genes now known to affect various traits and diseases of cats (much larger than the test battery Lyons was able to deploy on my students' cats just a few years ago). The information provided is obviously useful for knowing whether a cat has or may develop a genetic disease. In addition, breeders can use the information to decide which cats to breed, avoiding ones that might pass on undesirable traits. The tests are also useful for choosing cats that are carriers for desirable traits, eliminating cats that aren't carriers and thus can't pass on the beneficial allele.

In addition, an idea of a cat's ancestry can be provided, just as with humans. The UC Davis vet school lab uses a test devised by Lyons to indicate which of eight regions in the world a cat comes from, indicating the percentage of the cat's genome derived from each region (most American cats are predominantly Western European).

Several companies go an additional step, providing a breakdown of the contribution of different breeds to a cat's genetic makeup. In this, they parallel the information provided by dog genetic-testing companies. One dog-testing company, for example, reported on its website that Pepper was fifty-two percent Labrador Retriever and forty-eight percent Poodle. This more or less fifty-fifty split indicates that, most likely, Pepper's parents were a Lab and a Poodle. By contrast, Roxy's genetic mix was between eleven and fourteen percent—that is, about one eighth—of eight different breeds, suggesting that she had as great-grandparents a Siberian Husky, a Shiba Inu, a Chow, a Dachshund, and members of four other breeds.

Following the canine lead, the website of one cat genetic-testing company has an example in which a cat's ancestry was parsed into varying degrees of genetic contribution from Russian Blues, Ragdolls, American Shorthairs, Siberians, Siamese, Birmans, Persians, and others. Another company shows sample results indicating "we detected five breeds in Violet's DNA," namely American domestic (i.e., random-bred), Norwegian Forest Cat, Munchkin, LaPerm, and Siamese.

These breed breakdowns for cats are fundamentally misleading. Many dog breeds have been established for several centuries, some possibly much longer—enough time to develop breed-specific genetic markers. Moreover, the vast majority—perhaps ninety-five percent—of pet dogs in the United States today are mixed breeds descended from pedigreed dogs either recently or in their not-too-distant-past. So when a dog possesses a breed-specific allele, that means that it had an ancestor of that breed.

In contrast, all but a few cat breeds are very recent. Although several breeds have specific alleles—like the ones that identified Maine Coons and Burmese cats—most breeds do not. More importantly, the vast majority of cats today are not members of a breed, nor were most cats in the past. Thus the heritage of most cats alive today does not include pedigreed ancestors; most cats are not mixes of specific breeds like dogs.

How, then, do these companies come up with their estimates? The explanation is that not only is there geographic structuring to world-wide cat genetic variation, but there is also geographic structuring to breed origins. British Shorthairs, for example, are a European breed. If your cat is reported to have a large contribution of British Shorthair, what that means is that your cat has alleles found in cats from Europe, including British Shorthairs. Both your Fluffy and British Shorthairs come from the same stock. But it doesn't mean that Fluffy had a British Shorthair as an ancestor. In other words, don't place much confidence in what these companies tell you about the breed ancestry of your cat.*

THE FIELD OF EVOLUTIONARY ECOLOGY studies how evolutionary change through time adapts species to live in the world around them. It requires a synthesis of research on evolution in the past with investigation of how organisms are functioning in the present-day environments.

We've now got a solid understanding of the evolution of the domestic cat—where they've come from; how they've changed; their genetic, anatomical, and behavioral diversity. Now it's time to consider how they live their lives in the modern world.

*Case in point, the example I gave above for Violet's ancestry: the LaPerm is an exceedingly rare breed, and Munchkins are not very common. It seems highly unlikely that a cat would have members of both breeds as ancestors.

Sixteen

Pussy Cat, Pussy Cat, Where Have You Been?

Ever wonder what your cat's up to when she's outside?* The people in Shamley Green did. Located thirty-five miles southwest of London in the Surrey Hills Area of Outstanding Natural Beauty,† the picturesque hamlet has been home to the likes of Alfred Hitchcock, Richard Branson, and Eric Clapton's grandparents.‡

And cats. Lots of cats. So many that when the BBC convened an all-star cast of cat scientists to study what cats do "once they've left the cat flap," they chose Shamley Green for their filming location.

---

*Assuming, of course, that your cat goes outside, which is increasingly uncommon in the United States, though more common elsewhere.

†Yes, that's a real scenic-landscape designation in the UK.

‡Or maybe Clapton himself; the internet is confused on this point.

*The Secret Life of the Cat** is the story of a weeklong, high-tech investigation; fifty cats were monitored twenty-four seven, tracking their whereabouts and getting a cat's-eye view of life in the Green. The goal? To learn what cats do, and how so many felines can coexist in such a small place.

The documentary shows a project at a scale most scientists can only dream of: cameras set up all over town; an undercover surveillance crew stationed in a long, white panel truck full of television monitors and all manner of electronic gear; the town center commandeered, rechristened "Cat HQ," and turned into a cyber-command base with computer monitors flashing the latest updates, technicians chattering, staffers zipping back and forth. If only real field studies were as glamorous and exciting as the one-week project the BBC created for their show.

The documentary also features three of the world's leading ailurologists. Periodically, cat-behavior experts John Bradshaw and Sarah Ellis present mini-lectures interpreting what was seen on the latest video; in one long scene, Alan Wilson of London's Royal Veterinary College explains the technological wizardry that made the entire project possible.

But the heart of the story was the villagers and their feline companions. At the start of the film, a hundred or so locals pack into the community hall to hear the plan: the researchers wanted to put GPS units on the collars of fifty cats to track their movements throughout the village for a week. A subset of the focal felines—those deemed to be showing the most interesting activity patterns—would get a miniature spy camera added to their collar's payload to provide a cat-o-centric perspective on their exploits during their rambles.

---

*For those looking up the show, beware that the title *The Secret Life of the Cat* has been used countless times in books, videos, articles, and other media.

The hoped-for fifty cats (or at least their owners) quickly were on board. Collars were distributed, people went home, and then, as the narrator intoned in his English accent, "With the technology in place, it's now all down to the cats."

After a rainy start, the fun began. The weather cleared, out went the cats. Walking through yards, jumping up and over fences, walking down driveways, through gardens. The GPS units recorded the information—accurate to within inches—indicating where the cats had gone on their travels. Twenty-four hours in, the results were already fascinating. Projected onto big screens in the HQ, the path of each cat was displayed in different colors. Most of the cats—Brutus, Molly, Ginger, Hermie—hadn't traveled far from home. But Sooty, an ex-farm cat living on the village's edge, was more adventurous and had logged two miles in her peregrinations.

The cat-cam videos were just as interesting. Cats on the prowl during the day and at night, trespassing into other territories, hissing at each other, watching a fox cross a road, confidently going through the cat door in another house to nosh from a neighbor's bowl.

Great theater. And highly popular. The documentary was watched by five million people in England when it first aired in the summer of 2013, three times the usual viewership for the network's science programs. So successful, in fact, that the BBC produced a three-part sequel, *Cat Watch 2014*, the next year.

Secret Life's success highlighted not only the public's great thirst for information about cats, but also the paucity of scientific knowledge about the natural history of the world's most popular pet. At the time the show was filmed, there were few scientific data about where pet housecats go and what they do when they're out and about. That was about to change.

. . .

THE REASON THAT WE KNEW so little about the outdoor activities of pet housecats is simple: the secretive behavior of cats is legendary. If you've ever tried to follow a cat around outside, you've probably had no more success than I have. A "what are you doing?" backward glance is soon followed by a more prolonged glare. If still you persist, Luna quickly reminds you who's the boss by heading to the nearest dense shrubbery—festooned with thorns whenever possible—and disappears from sight. The mystery of the cat endures.

Biologists have a trick for studying hard-to-follow species, the same ploy used by spies to track their quarry: put a tracker on them, usually in the form of a transmitter that broadcasts a radio signal from a collar around the animal's neck. You've probably seen nature documentaries with elephants or lions wearing these necklaces. Some animals are too small or don't have necks (like snakes), in which case the transmitters are attached elsewhere on the body or implanted surgically.

The advent of radio-tracking in the middle of the last century revolutionized our understanding of animal movements, and the approach has been used on animals ranging in size and lifestyle from insects to bats to whales. But the method comes with a major disadvantage: data collection is very time- and labor-intensive. Researchers have to go out into the field and get close enough to be able to detect and localize the signal. How close depends on the size of the transmitter, which in turn depends on the size of the animal—bigger critters can carry larger equipment. Conversely, tracking small animals has the advantage that they tend not to move as far as larger ones (except for those that fly), so the researcher can travel on foot, starting where the animal was last detected. For more mobile species, on the other hand, cars and even airplanes are needed to cover enough ground when searching for the radio signal.

A study in Albany, New York, is an example of this sort of research. Scientists monitored eleven housecats over the course of the summer in a suburban neighborhood near a small nature preserve. Eight households bordering the reserve had volunteered to join the study; their cats were fitted with collars bearing two-ounce radio transmitters with six-inch-long antennas for broadcasting the signal.

This was before the advent of GPS units, and tracking was no easy feat. To locate the cats, team members would drive to the neighborhood, get out of their car, don their headphones, and turn on the radio receiver. Each cat's transmitter produced signals at a different frequency. The researchers would tune the receiver to each frequency, listening for the telltale beeping from the collar. By rotating the antenna one way then the other, the beeping would get louder or softer. Then, they'd walk or drive in the direction in which the beeping was loudest until the noise was deafening (or at least it would have been if there weren't a volume knob). At that point, the cat was right in front of them.

This approach is particularly effective when you're already close to the cat from the beginning. If the cat is not initially nearby, you have to determine the direction in which the beeping is loudest and then draw a line on a map in the direction of the sound from where you're standing. Then you move to another location some distance away and repeat the procedure, plotting another line. Where those two lines intersect is where the cat should be (assuming that she didn't move while you were traveling from the first point to the second, which would complicate matters, especially if the cat moved a large distance).*

The researchers were surprised at how unadventurous the cats

*This procedure is called "triangulation," because the three points—the two observer points and the cat—form the corners of a triangle.

were. Most didn't wander far. On average, they only covered an acre and a half in their meanderings during the summer (slightly more than a football field), leaving their own yard to visit three neighboring properties and rarely entering the forest preserve.

Tracking was a time-consuming process, generally only yielding one location point per cat per day; what the cats did during the remainder of the day was missed. Moreover, most of the cats spent much of their time inside, which explains why there were only about fifteen outside locations for most cats (the exception being Orion, the youngest cat in the study, who was located outside fifty-six times!).

Happily, technological advances have been a game changer in animal tracking. Radio transmitters have been replaced by GPS units that receive positional information from satellites. As you know from your own navigational devices (be they in cars or your cell phone), these systems update themselves nearly instantaneously, providing a steady stream of locality data. Instead of one location point per day, these units can provide one data point per second!*

Maybe it was the additional data points or perhaps it was the village's more bucolic setting, but what the BBC show revealed is that the lives of Shamley Green's grimalkins are charmingly varied. Certainly, there were some stay-at-homes like the Albany Eleven. Rosie, for example, was a fawn-colored British Shorthair who spent a lot of time outside, but mostly in her own backyard, covering less than a half acre in her travels. Another homebody, Coco, didn't cover much ground either, rarely going more than a few houses away from her home.

At the other extreme, the most traveled of the show's felines was Sooty, black with glowing yellow eyes, who lived in a house backing

*Most GPS trackers record data less frequently to keep from quickly running down their batteries.

up to countryside, providing wide-open spaces in which to roam. And roam she did, traversing more than seven acres over the week of filming.

This variation in prowling behavior even extended to cats within the same household. In the only six-cat residence in the study, five of the felines—Duffy, Daisy, Pumpkin, Ralph, and Coco—barely left their own property, whereas a black Exotic* named Patch wandered widely, covering about five times as much ground.

On the other hand, some cats were very similar in their excursions. For example, two cats that lived in houses across the street from each other covered almost identical real estate, rambling through the same three-acre section of their neighborhood. Despite this shared domain, the time stamp on the GPS data revealed another surprise: the cats managed to avoid each other almost completely by being active at different times of the day.

Even as the *Secret Life* scientists developed a very sophisticated GPS system in 2013, much cheaper, simpler versions were becoming commercially available. Ever since, people have been buying them, placing them on their cats' collars, and monitoring their pets' activities.

Today the market is awash with such cat-tracking products. For example, as I was researching this book, *Pet Life Today* posted an article on "The 25 Best GPS Cat Trackers & Collars of 2019." For the more discriminating, or those on a tighter budget, *Buskers Cat* settled for "The 9 Best Cat Trackers of 2019." Despite—or perhaps because of—this cornucopia, the two sites were in almost complete disagreement about which was the feline tracker extraordinaire. I had no choice but to find out for myself.

---

*Exotics are a breed of cat, a short-haired version of the Persian.

. . .

WINSTON IS A BIG, tall cat, seventeen pounds of mostly muscle. Plus, he's white, with large gray-black patches. At a distance, you might mistake him for a tiny Holstein. Gleaming like a mega-marshmallow, he's hard to miss when he's anywhere in our backyard. But most of the time when he's outside, he's nowhere to be seen. Where is he, and what's he getting up to?

Winnie's not a fan of having the breakaway safety collar placed around his neck, but once it's snapped on, even with the small GPS transmitter hanging from it, he doesn't seem to mind. Off he goes into the neighborhood wilderness, all the while the GPS unit talking to my cell phone, updating his current location every ten seconds and keeping track of where he's been.

Winston wearing a kitty camera.

I'm a hopeless procrastinator when I'm trying to write, and so from time to time I'd stop typing, pick up my phone, and open the tracking app to learn Winston's whereabouts. It was always a particular thrill to look at the little screen and see the blue catface dot on the satellite map indicating he was in the backyard, perhaps on the left side by the hydrangeas. I'd look up, turn my head to look out the window, and as if on cue, there he was, my miniature cow of a cat, walking through the garden.

Other times, the catface would indicate that he'd left our yard and was wandering elsewhere in the neighborhood. Occasionally overcome with curiosity or dubious of modern technology, I decided to ground truth the accuracy of the unit by walking to the spot where he was supposed to be. And sure enough, every time, there he was, lying under a bush or sauntering through a yard, sometimes clearly not so happy to see me.

As well as pinpointing Winston's current location, the app also showed his perambulations. Again superimposed on a satellite map, the "history" page featured a line tracing his travels over the course of the day.

I had no idea Winston covered so much ground! Winnie spent most of his time patrolling our backyard, king of all he surveyed—property boundaries apply to felines as well as humans, it seems. But once or several times a day he would venture beyond our jurisdiction. Over the fence into the grumpy old man's backyard, across the shallow creek into the quiet subdivision behind us, then back into the other grumpy old guy's side yard. Crossing the subdivision roads didn't bother me (especially for a very visible white cat), but I was disturbed when one day his track crossed the much busier road on the other side of our house.

On a typical day, Winston roams across about four acres of land, visiting four to five of our neighbors. Of course, he doesn't follow the same route every day. All told, his "home range"—the zoological term for the area an animal visits over a period of time—is perhaps eight acres, encompassing the fifteen or so properties closest to our own.

Much as I think of my feline companions as exceptional in every way, Winston's ramblings are completely typical for pet cats. He's moving around, occasionally meeting cats and other animals, hunting, and doing a little catnapping. I had an inkling from watching *Secret Life,* but now I know a lot more about why such information is important to our understanding of housecat biology thanks to a remarkable citizen-science project based in Raleigh, North Carolina.

ROLAND KAYS IS THE HEAD of the Biodiversity Research Lab at the North Carolina Museum of Natural Science. A specialist on carnivorous mammals, the Michigan native has conducted pathbreaking work on the distribution of coyotes as well as on the man-eating lions of Tsavo and the social behavior of kinkajous (a long-tailed tropical relative of the raccoon). He is also famous for using new technology in creative ways, such as putting miniscule radio transmitters inside the acorn-like seeds of tropical trees to track their fate (large rodents bury them for a future meal, but then repeatedly come back, dig them up, and move them to a new spot), or using drones outfitted with thermal cameras to locate howler monkeys at night, glowing in the infrared spectrum against the cooler leaves of the rainforest canopy.

Kays was one of the first to radio-track housecats. Using radio collars left over from a student's work on skunks, he was the leader of the project in Albany that I've already described. Subsequently, he went on to man-eating lions, coyotes, and other studies. Arriving in Raleigh in 2011 to head the biodiversity lab, he didn't have any plans for further feline research. But then, a decade after his cat radio-tracking study, he learned about the existence of cheap commercial cat GPS units.

Still, cheap is not the same as free, and Kays already had a full plate of administrative, teaching, and research obligations. Where would the funding and time come from to run the project? And then a flash of genius: get people to track their own cats—citizen science in action—and have undergraduates at nearby North Carolina State University run the project!

Kays quickly learned there is no shortage of star students interested in getting wildlife research experience, even if—or especially when—the wildlife is none other than *Felis catus*. And so Cat Tracker (as the project was named) began. The goal was simple: find out how far pet cats travel when they're outside.

Knowledge of an animal's movement patterns—where it goes, how widely it ranges—is critical to understanding how it interacts with the environment. For example, cats are consummate predators, and there is great concern about their impact on the environment. Consequently, it matters a great deal whether a cat is staying in her backyard or wandering into a nearby nature reserve. Moreover, the farther a cat wanders, the more likely she is to get into trouble by crossing roads, encountering mean dogs (or worse, coyotes), or otherwise jeopardizing herself. And then, of course, there's just plain curiosity. My guess is that most, if not all, cat custodians wonder, like me, where their companions go when they're out and about. The question has evolutionary implications as well: Do housecats differ from feral cats in their wanderings, and how does *Felis catus* compare to its wild relatives?

The project's approach was simple: advertise widely and wait for people with outdoor cats to sign up. Participants—primarily in the Raleigh-Durham area of North Carolina and on Long Island, New York—obtained a cloth harness for their cats to wear. The harness was minimal: a collar around the neck, another collar that fit around the cat's midsection, and a piece of fabric that sat on the kitty's back and connected the two collars. Attached to this connector fabric was the GPS unit, the size of a small Halloween candy bar and encased in a soft plastic holder.

A cat wearing a cat tracker.

I've put such harnesses on my cats, and after a few minutes of annoyance, they ignored them completely and went about their business. In the North Carolina group's studies, only a small proportion of the cats would not tolerate wearing them.

Cats wore their mini-backpacks for about a week, the GPS unit recording their whereabouts. Preliminary data showed that five days of data were enough to get a good estimate of the area roamed by a particular cat. Because there were a limited number of units (they weren't that cheap!), after a week the cats were thanked for their service, the harnesses removed, the GPS units returned to the lab, and the data downloaded. Then the tracker would get sent to a new cat for another weeklong tour of duty.

Just as I did with Winston and the Shamley Green team did with the village's cats, the researchers were able to visualize each cat's journeys over the course of the week. Plotting the locations and paths on satellite map images, they were able to calculate the cats' home ranges, as well as learn where they'd been: how many people's yards, how many roads crossed, how many forests entered, and other relevant details.

There were computer glitches, of course, but more challenging were data points that were correct but indicated something out of the ordinary. One cat, for example, suddenly moved a great distance at a speed of thirty-five miles per hour, much farther and faster than housecats travel. A quick call to the cat's owner revealed that the cat had taken a trip to the vet while wearing the GPS unit; the owner had meant to report the outing to the researchers but had forgotten to do so.

In another case, the location points indicated that a cat crossed a river, then a few hours later crossed back. Checking with the owner revealed that the data were correct; the Maine Coon had traveled over the frozen river in the middle of the winter (the cat liked to swim, so who knows what would have been revealed had he been tracked during the summer? Probably the researchers would have learned that the GPS units aren't waterproof).

And then there was the cat that repeatedly came home smelling of cigarette smoke. The woman who owned the cat suspected infidelity, and the tracker proved her right—the cat was two-timing her at another house. Such unfaithfulness was common; when confronted with tracking data, a number of homeowners fessed up to feeding a neighbor's cat.

The trackers also solved the mystery of cats that disappeared for several days at a time. One cat made the mistake of doing so while he was being tracked. His owner now knows to drive down to a nearby business park to pick him up whenever he goes missing. In another case, a cat owner who had desperately searched the entire neighborhood for his missing puss learned the cat had been in his house the whole time.

A visit to the Cat Tracker website shows the ranging pattern of some of the project's favorite cats. Catniss Everdeen, for example, is a beautiful blue-eyed, long-haired youngster with Siamese points. Barely more than a kitten, the one-year-old lived in a house on a two-lane road in Durham. Most of her activities centered around her house and the small forested lot out back. However, Catniss did make several visits to the apartment complexes on both sides of the house, as well as crossing the road three times, once going more than 150 yards to an industrial parking lot. The total area she ranged was approximately four acres.

Little is an eight-year-old orange tabby male that lives less than a mile away from Catniss on another two-lane road in a tree-shrouded neighborhood. His activities, too, were centered around home, but he took longer forays, the farthest being more than a third of a mile to a wooded tract many streets away. His overall range was much larger than Catniss's—about thirteen acres.

At the other extreme, Shadow is a lovely brown cat with gorgeous yellow eyes living in Greenwich, Connecticut. The large number of tracks on his map indicates he's a very active three-year-old, but he doesn't go very far. Most of his travels are to the houses on either side of his, with the occasional jaunt to another nearby house. His longest movement was less than the span of a football field, and overall he had a home range of less than an acre.

Shortly after Kays started the Cat Tracker project, teams in Australia, New Zealand, and England joined the project, making Cat Tracker a truly global endeavor. Across the four countries, more than nine hundred cats participated in the study.

Going into the study, the researchers made predictions about how the cats would vary in how far they roamed. Rural cats might wander farther than city slickers. In places with large predators—specifically, coyotes—cats might stick closer to the safety of home. Male housecats might travel more than females, as they do in many other mammal species. There were many hypotheses to test.

The main message from the data was clear. Just as Kays found fifteen years earlier in Albany, the vast majority of cats didn't go very far. The average feline traversed only thirteen acres, and that number was reduced to nine when three cats with exceptionally large ranges were excluded.

But let's talk about those three exceptions. Far and away the most wide-ranging cat was Penny, a spayed one-year-old kitty from the suburbs of Wellington, New Zealand, who spent most of her time outside. Penny lived in a house that backed up to a hilly undeveloped area, and she roamed widely all over it, covering a range of more than two thousand acres—three square miles—nearly four times as large as the range of the next most wide-ranging cat in the study.

That next cat was Max, a five-year-old neutered male from the far southwest of England whose range was 550 acres, almost a full square mile. Max traveled like no other cat in the study, walking along a road for just over a mile from the village of St. Newlyn East to Trevilson and back, twice during the six days he was tracked. If I had to speculate, I'd guess that at one time, Max lived in one village and then he and his entourage moved to the other. Cats are renowned for their tendency to find their way back to former residences* (as far as I know, though, those stories don't involve a cat going back and forth between new and old homes). Max's journey will have to remain a mystery.

Rounding out the Exceptional Three was Blue Nelson (not to be confused with my brown Nelson!) from the northern tip of the South Island of New Zealand. Not allowed outside at night, BN made up for his nocturnal detention during the day, roaming widely amid large fields surrounding the farm on which he lived, covering almost as many acres as Max.

These three wanderers notwithstanding, most cats were much less adventurous—more than half the cats had ranges smaller than two and a half acres; ninety-three percent used home ranges smaller than twenty-five acres.

By contrast, thanks to radio-tracking studies, we know that feral housecats range over much larger areas: on average, four hundred acres in Illinois farmland, twelve hundred acres in the Galápagos Islands, a whopping five thousand acres (eight square miles!) in the Australian Outback. Small, wild feline species are similarly wide-ranging. European wildcats often have home ranges of twenty-five hundred acres or more. Other small cat species like ocelots and Geoffroy's cats have comparably expansive ranges.

Why do pet cats have such small ranges? The answer is obvious—they're getting fed at home and have no need to travel far and wide to find their next meal. Also, most housecats are neutered, so there's no urge to search for a mate. Whatever psychological benefits pet cats get

*The internet is full of stories of cats finding their way to former homes. Pet websites provide advice about how to keep cats from doing so when you move. Specifically: after moving to a new residence, keep your cats inside for a long time before letting them out for the first time. I'm unaware of any scientific research on this topic.

from being outside, they're generally satisfied by wandering over a much smaller space than felines that need to look for food or sex. The comparison between feral cats and other felines species, however, suggests that *Felis catus* has not evolved to be less wide-ranging than their wild brethren—they're just as wide-ranging when circumstances warrant.

The Cat Tracker team expected to find geographic differences in cat behavior. In the United States, for example, the widespread occurrence of coyotes might inhibit cats from wandering. Also, owners in some countries might feed their cats more or less; cultural differences of all sorts (landscape layouts, car speeds, presence of dogs) might have an impact.

Surprisingly, the differences among countries were minor. The home ranges of cats in the US, New Zealand, and the UK were all about the same size, and nearly double the home range of cats Down Under. Why Aussie cats range less is not known, but regardless, the major finding is that pet cats everywhere don't generally travel great distances. Or, as Kays summarized more bluntly, "Pet cats are universally lazy." Of course, the four countries included in the study are similar in many respects. Investigating whether these findings extend to Latin America, Asia, or Africa would be an obvious next step.

One other notable difference between countries was discovered. In Australia, most cats covered much more ground during the night than during the day, whereas in New Zealand, some roamed more at night, but an equal number roamed more during the day (data on day versus night roaming were not reported for the other two countries). The explanation for this discrepancy is unknown.

An amusing side note is that in both these countries, many owners said their cats were only allowed outside during the day, but the trackers revealed that many of these cats (twenty percent in New Zealand

and thirty-nine percent in Australia) did, indeed, go out at night and traveled extensively. At face value, it's hard to understand how a cat not allowed to go outside is nonetheless getting outside, and the statement of an Australian owner of one of these cats that "I know our cat is very sneaky and if he wants to get out at night he will find a way!" doesn't really clarify matters.

It never pays to underestimate the guile of a feline. My dad's cat Aten learned how to jump up and pull down the lever-style handle on the front door to get out. "Smartest cat in the world," my dad claimed, but Aten wasn't smart enough to push the door closed when he came back in, and quickly the mystery of the open front door in the middle of the night was solved.

Most likely, the owners of these cats aren't really trying to prevent them from going outside at night, they're just not observing the cats doing so. As one resident of Shamley Green explained about a similar surprise finding during that project, "When we go to bed he is asleep on our bed, when we get up at six a.m. he is still asleep on our bed, yet overnight he has been miles! I was sure I had a pretty good grasp on what my cats get up to, but that was a real revelation." Still, that begs the question: Are the cats being intentionally devious by slipping out unnoticed? Clearly a topic deserving further research!

Other findings confirmed many of the researchers' initial predictions. Males covered more ground than females, and the few "intact" cats had substantially larger ranges than the neutered ones. Rural cats wandered over greater area than city slickers. On the other hand, even though many cat breeds are known to be very sedentary, pedigreed and unpedigreed cats did not differ in their extent of roaming.

I've focused so far on the size of the area over which cats roamed, but of course in this era of satellite maps and geographic information systems, location data provide much more information than just position coordinates. For example, concerned about whether your Priscilla is in danger of being run over? Maybe you should be—the average cat crossed roads four and a half times during the few days she was tracked.*

Another important question concerns the types of habitats the cats are using. Kays's Albany cats primarily limited their forays to backyards and other human-modified places. Would the same be true elsewhere? For the most part, yes. Across the four countries, three quarters of the cats spent almost all their time in such "disturbed" environments. Nonetheless, a minority of the cats bucked this trend: one cat in ten spent most of her time in natural habitats such as forests and wetlands.

THE CAT TRACKER PROJECT and similar studies of different scale and time frames have gone a long way to illuminate the outdoor activ-

---

*The internet is full of claims that according to a National Traffic Safety Commission study, 5.4 million cats are hit by cars each year in the United States, ninety-seven percent of which die. However, I've searched extensively and can't find any evidence for such a study. I've queried some of the people who've posted that information, and they can't provide documentation either.

More reliable data come from an English study that surveyed cat owners periodically to monitor the lives of their cats. Among the more than twelve hundred young cats that were allowed to go outside, four percent were struck by a car over a twelve-month period. Most of the cats were killed in the accident. Four percent may not seem like a lot, but projected over the life span of a cat, the odds of getting hit are fairly high (although older cats are probably both wiser and less adventurous, and thus probably are run over less frequently). Incidentally, four percent of the number of pet cats allowed to go outside in the United States is vastly fewer than 5.4 million cats. The number of pet cats in the US is somewhere between fifty and one hundred million, of which only thirty percent go outside. Four percent of that number is around one hundred thousand road-killed cats per year, still way too many.

ities of our cats. The findings have important implications for cat safety and environmental impact, as we'll explore shortly.

Yet, fabulous as these cat-tracking studies are, they don't directly shed light on perhaps the most intriguing question of all: what are our cats *doing* while they prowl around outside?

Seventeen

Lights, Kittycams, Inaction!

I've already noted how hard it is to observe roaming cats. But there's another way to figure out what they're doing, one that lets us see the world from the cat's point of view. Just as technological advances allow much easier and more detailed localization of cats, progress in remote videography has given us the cats' own perspective on what they're up to when they're on the go.*

It all started when an aspiring graduate student interested in the welfare of stray cats teamed up with a veterinarian turned professor at the University of Georgia and the techno-wizards at the National Geographic Society. Having led several semester-abroad programs in

*Keep in mind that cat and human vision differ somewhat. Cats are much better at seeing in the dark than we are, but that comes at a cost of not having as acute vision during the day. In addition, their ability to focus is not as good as ours, but they are very good at tracking moving objects. Also, cats are green-red color-blind.

Tanzania, Kerrie Anne Loyd wanted to study large African carnivores for her PhD. But before she went down that road, she became obsessed with the cats roaming her own neighborhood in Atlanta. They weren't healthy and were killing a lot of birds and other wildlife. The final straw was the day she was sitting on her front porch and a neighbor's cat walked out from inside her house, Loyd's pet cockatiel in her mouth!

Loyd shucked her Africa plans and contacted Sonia Hernandez, a professor of wildlife epidemiology at UGA. Hernandez, who had been a practicing vet before getting her PhD, was also concerned about stray cats, both their welfare and their effect on wildlife. Loyd's timing was fortuitous because Hernandez had already been in contact with the National Geographic Society about developing a small camera that cats could wear to record where they went and what they did. And thus kittycam science was born.

The project faced two obstacles: finding people willing to let their cats be the guinea pigs for the project and figuring out how to make a cat-borne camera. The first part turned out to be easy: owners were delighted to participate because they, too, wanted to know what Jasper did on his forays outside.* Developing wearable feline cameras was a bigger challenge.

The National Geographic Society started its Crittercam™ program in 1989 to develop cameras that could be attached to or worn by animals, revealing their daily activities from their point of view. Such cameras were designed to study species that are secretive, live in remote places, or are too dangerous to observe at close distances.

Initially, the cameras were large and clunky. Composed of a hand-

---

*Though not all were happy with what they learned. One woman was so appalled to learn that her cat had caught and killed a bird that she almost withdrew from the project. Instead, Loyd convinced her to buy a cat bib that makes it harder for cats to catch birds.

held VHS camcorder encased in a tube-shaped protective casing, they could only be used on animals large enough to carry the contraption's weight (the rule of thumb in zoological studies is that devices should be put on animals only if they weigh no more than three to five percent of the animal's overall weight or, ideally, less). Suction-cupped to the back of a whale, clamped to a shark's fin, strapped on the back of a king penguin, Crittercams provided unprecedented views of the lives of animals, showing us things "you don't know you don't know," to quote Crittercam creator Greg Marshall.

Kyler Abernathy was an early enthusiast. As a master's student studying the biology of the Hawaiian monk seal, Abernathy was present when a Nat Geo team arrived to deploy a camera on a seal. At first, he was skeptical: it seemed more like a gimmick than a useful scientific tool, and he feared that the seal would be bothered by the camera and wouldn't behave naturally.

Then he saw the first video footage and it changed his world. "It overturned everything in my head about how they lived their lives in the ocean. All of my mental imagery was wrong. I realized, holy cow, I was asking all the wrong questions. I was thinking the wrong things. It turned me around about the system."* When he completed his master's degree in 1998, he got a job working on Crittercam development and never left.

The potential of Crittercams is so great that Nat Geo's Exploration Technology Lab, which Abernathy now heads, is always inundated with many more requests for help from researchers than they can possibly accept. As a result, they have to choose which projects to

*Specifically, he thought that the seals captured fish by swimming in the nearby shallow coral reefs, but the camera showed that, quite the contrary, the seals swam past the reefs and dove deep into the ocean to find their prey.

work on. Deciding can be difficult. The team weighs the scientific and conservation importance of different species, as well as the feasibility and cost; some projects require too much new engineering. Occasionally, they choose a species because its natural history poses an exciting new challenge.

Hernandez's timing couldn't have been better. Prior to that time, the technology wasn't in place to make a camera small enough for a housecat, but by 2010, that had changed. Nat Geo agreed to take on the challenge of developing a high-tech kittycam, small enough for a cat and with an infrared light for nighttime filming.

The result was a box three inches long, two inches high, and an inch deep that weighed about three ounces (equal to two percent of a ten-pound cat's body weight). The lithium-ion battery could record for ten to twelve hours; videos were stored on a microSD card like those used in point-and-shoot cameras.

This description doesn't fully capture the serendipity of the design process. As well as the technical elements—size, electronics, optics—the lab had to devise a container that was rugged and elements-proof, yet easy to open to change the batteries and video card. Frankly, the lab's engineers were stumped—they'd never had to grapple with these issues on such a small scale.

Then the wife of Abernathy's boss—who was aware of the issue—happened to notice the perfect container while shopping at CVS: a travel tampon case. Two oblong pieces of plastic, one of which snugly slides over the other to form a carrying case. Just the right size, lightweight, and easy to open and close. Somewhat self-consciously, Abernathy repeatedly visited local drug stores, buying their entire stock. The staff behind the counters were pros and never said a thing.

The camera plus batteries fit inside beautifully, holes were cut for the lens and light source, and the kittycams were ready to go. Abernathy and team knew the containers weren't perfectly sealed, despite their best efforts. This would never work for most species, for which the cameras have to be designed to handle all conditions.

But these weren't most Crittercam critters. When it's pouring, any cat with any sense would stay inside. And for cats without sense, Abernathy's team asked the owners to not put the cameras on their cats when the forecast called for rain.

A cat wearing a National Geographic kittycam.

Still, there were issues. Several cats liked to play in streams. Another cat had the habit when drinking from his water bowl of supping from the far edge, stretching his head and neck across the entire dish and, in the process, dunking the kittycam. Several other cameras returned to the lab because they had stopped working were found to be full of kitty litter. Overall, though, the cameras functioned quite well.

In an average deployment, the cats wore the camera for five hours. Fifty-five cats participated in the study, and each cat was adorned with a camera seven to ten times. Do the math, and you'll come up with an average of nearly forty hours of video per cat, more than two thousand hours in total.

That sounds like a lot of footage to review, but the job wasn't quite that enormous because the cameras were designed to stop recording during periods of inactivity, and the cats, being cats, spent about two thirds of the time resting or sleeping. Throw out the time the cats spent indoors while wearing the cameras, through which Loyd could fast-forward, and there were only about five hundred hours of video footage to watch. Better, but that's still equivalent to binge-watching all eight seasons (seventy-three episodes) of *Game of Thrones* . . . seven times. All 201 episodes of *The Office* five times. You get the idea.

The UGA Kitty Cams project has posted a gallery of greatest hits videos, seventeen in all, that give a taste of what it's like to review the footage. The first thing you need to do is orient yourself. The camera is on the cat collar, hanging under the neck. As the cat walks, the camera bobs up and down, the perspective a bit stomach turning for those prone to motion sickness. A rounded object periodically intrudes from the top of the frame, its identity puzzling until the appearance of whiskers on either side make clear that it's the cat's jaw.

The UGA website highlights the magic of the kittycam. The three-minute, fifty-two-second mini-saga "Bringing Home a Chipmunk" begins with a chaotic view of a lawn, seesawing like a violent sea as the cat walks rapidly. Swinging in and out of view at the top of the screen are two hindlegs, then legs and a tail, then a head, confirming the rodent's identity. The view is of the bottom of the chipper—legs, tail, belly, head—mentally you fill in that the cat is grasping the rodent by the scruff of its neck or back. Bailey* is walking across a lawn, but

*The UGA website didn't provide the names of the cats, so we'll use the most popular cat name in Georgia, according to a pet insurance company survey.

when he comes to a driveway with some parked cars, he breaks into a run, scenery and chipper swaying wildly back and forth, thudding footsteps sounding like a galloping horse (at least, I think that's what the sound is, though perhaps it's the camera rubbing against the collar?). When Bailey gets to the split-level house, he starts mewing, bounds upstairs to the main-floor level, and seems to be circling the house looking for a way in. Finally, he comes to a glass door, crying plaintively. Someone inside walks up, then turns away (the owner's voice can be heard as a muffled sound, probably something to the effect of "I'm not letting you in with that chipmunk!"). Suddenly, the striped mouse starts moving its head—he's alive!—looking around, then straight at the camera. The owner comes back to the door. All of a sudden, the chipper is gone. I replayed that fraction of a second countless times trying to ascertain exactly what happened—I'm pretty sure the cat either dropped the chipper or it wriggled free. In any case, it's gone, but Bailey doesn't seem to care, looking forward at his owner's shins, expectantly waiting for the door to open. The scene fades to black.

Other cat's-eye videos include watching a bunch of free-range chickens on the loose in a backyard; jumping to the top of a fence, then down to the backyard on the other side for a look-around; meeting another friendly cat; growling loudly at a dog from underneath a car; and a two-and-a-half-minute walk through a big yellow house's pine cone–strewn front yard.

One of my favorite clips illustrates how hard it is to figure out what happened. The fifty-five second clip titled—inaccurately, I believe—"Fighting Off an Opossum" begins with the viewer looking up at an opossum perched on the top of the railing surrounding a patio deck. The background blackness tells us it's nighttime, as does the opossum's bright eyeshine, reflecting the infrared light put out by the kittycam. The opossum begins by scratching its side with its left hindleg, then takes a few steps forward on the flat top of the railing. It stops and gives its chin a thorough scratching with its right hindleg. Then it continues walking along the top of the railing. The perspective shows us the opossum must be walking directly above the cat, which is looking straight up and rotating his head left to right as the marsupial passes overhead. At this point, the railing takes a right-hand turn along descending stairs at the side of the house, in front of a brightly lit window. As the opossum reaches the turn, it is lost from sight in the bright glow from the window; the cat continues to turn his head to the right, following the trajectory of the opossum, which is no longer on the top of the railing, nor in sight at all. All of a sudden, the opossum is directly in front of the cat—how it got there is not clear. There is a scuffling noise as the cat moves, though no obvious hissing or growling. The screen goes white, indicating that the opossum is directly in front of the cat, too close for focus and flooding the image with reflected light from the camera. Then we see the opossum again, then more whiteout, then the opossum crawling back between the slats of the railing, down the stairs, and into the darkness.

Was this an actual fight? I don't think so. Surely there would have been some hissing or growling from one of the would-be combatants, but none is heard. I think the cat just lost sight of the opossum, which suddenly appeared two feet or so away, possibly not having seen the cat either. Both were surprised; the opossum moved off. Peace prevailed.*

---

*I need to put my cards on the table here: I'm a card-carrying member of the Opossum Fan Club and feel that these fascinating animals, the only marsupials in North America, get a bum rap (marsupials, such as kangaroos and koalas, are mammals that raise their young in pouches). Sure, to the uninitiated they may look like overgrown rats, and their fifty fearsome teeth (the most of any North American mammal!) are scary, yet in reality, they are harmless omnivores that do us a service by scavenging carrion and eating pests. Moreover, in these days of bad

The UGA videos are spectacular, providing fascinating vignettes of the lives of the cats they studied. Still, these are the highlights of the project, and all together they only add up to a half hour of footage. My guess is that the videos must have been full of a lot of dull nothingness, the cat turning its head one way, then another, or walking around doing and seeing little. Watching five hundred hours of mostly that sounds tedious, but Loyd was upbeat, saying that it wasn't so bad and that's what graduate students do to get their doctorates.

To find out for myself, I gave cat voyeurism a try by purchasing my own kitty camera. Having watched a number of videos of Winston and Jane traversing our backyard and the neighborhood, I have heightened respect for Loyd. To put it simply, the experience of watching these videos is an exercise in boredom and frustration.

First, the boredom. Big surprise: cats do a lot of sitting around, seemingly doing nothing. Like the Nat Geo kittycam, commercial models are supposed to shut down when nothing's happening. And sometimes they do. But it turns out that you get what you pay for—when you plunk down only a hundred bucks, you don't get the same reliability you'd get from Nat Geo's one-grand version.* So there's a lot of video of the same, unchanging scene when the camera fails to stop recording. Plus, cats spend a lot of time looking around, moving their heads from side to side, perhaps responding to smells and sounds we can't detect. Even when the motion sensors are working, this activity is enough to keep the video rolling.

Watching just a few of these videos is mind numbing; looking at

---

news about disappearing wildlife, opossums are a great success story, one of the few species thriving in human environments. So next time an opossum waddles by in your backyard or crosses the road in front of you, give the little fellow a tip of your hat and leave him be.

*That's an estimate, as Nat Geo doesn't sell them.

bushes, trees, driveways, and jungle gyms gets old pretty quickly. I can only imagine the monotony of a five-hundred-hour felineapalooza. But Loyd has it right. That's science. A common caricature involves a scientist in a lab coat at an equation-covered blackboard yelling, "Eureka! I've figured it out!" In reality, though, most science involves hard work, with a lot of repetitive monotony: mixing chemicals together time and time again to see what happens, measuring thousands of plants to see how chemicals affect their growth, watching mice run on a wheel for hours to test ideas about the effect of different diets or amount of daylight. The ideas being tested are interesting; the actual data collection, not so much. Even watching elephants for long periods can be tiresome!

Now, for my particular frustration. Every step Winston or Jane took, the camera bobbed up, then down. Much more than with the Nat Geo kittycams. Maybe it was the lighter weight of the commercial model (0.3 versus 3.0 ounces) or maybe they're just not as well made.* But whatever the reason, whenever the cats were on the move, the images rapidly oscillated, making it hard to tell where the cats were, much less what they were doing. And that's when the cats were walking. When they ran (as when Winston crossed the small road behind our house—good boy, Winnie!), the images were a complete blur. Sometimes, I could figure out events by the context. Jane walks up to a fence, an uninterpretable montage of light and dark, and then she's surveying the neighbor's yard from a high vantage. Clearly, she had jumped to the top of the fence. But in other cases, I had no idea what just happened.

Of course, that wasn't always the case. When Winston retched up

---

*I was very disappointed that the first unit I bought stopped working after a few months of modest use. I bought another and the same thing happened. Checking reviews online, this seems like a common issue.

a hairball—one of the most exciting moments of my mini-study—I figured out what was happening from the forward-rocking oscillation, the all-too-familiar gagging sounds, and the object that sailed through the camera's view to the ground after the last gag. Poor boy (but at least he wasn't on the nice rug this time!).

I've watched a lot more lizards—either live or on videotape—than cats (it's my day job, after all). If anything, watching lizards is even more monotonous, because lizards do truly spend most of their lives doing nothing. Still, those few moments of saurian excitement—a lizard running down the tree to grab a passing cockroach, males engaged in a territorial encounter, or something completely unexpected, like a lizard eating a red berry—are enough to make up for the ennui.

And the same is true for cat self-videos. The highlights other than the hairball:

- Jane going on a much longer trek than I ever expected, ending up by a red wheelbarrow that took me a long time to find when I tried to retrace her steps across several neighbors' yards (difficult as I tried to look inconspicuous while peering into people's side and backyards).
- Winston entering our house and hissing at Nelson as he came running up.
- Winston coming to the glass sliding door multiple times, me getting up from my comfy chair to let him in (an out-of-body experience to watch myself from Winston's perspective), and—psych!—he fools me again, turning around and heading back to the garden, almost as annoying on videotape as it is in real life.

No action shots of Jane stalking a rabbit nor Winnie confronting a raccoon, but then again, I only watched about ten hours in total.

But enough of amateur hour. Let's go back to the real cat researcher. What did Loyd discover? Just like Winston and Jane, housecats in Athens, Georgia, spend most of their time outside doing nothing. "Many of them did a lot of lounging, just sitting on the porch, waiting for their owners to come home," Loyd said. All told, the cats spent about three quarters of their outside time sitting still, either sleeping, resting, or grooming.

Nonetheless, when they were out and about, the Athenians got into all kinds of mischief. For example, one cat crossed a wide street to a storm drain and ducked into the narrow horizontal slit in the curb. Without hesitating, the cat then jumped down several feet to the bottom of the brick-enclosed space and walked through a corrugated-iron culvert for perhaps fifty feet until the drain narrowed and took a turn to the right. At that point, the cat stopped, looked back, looked at the opening again, perhaps thinking, "Hmm, maybe that's not such a good idea," then turned around, retraced his steps, jumped back up to the entrance and reemerged on the street.

This event seemed particularly dangerous—what if a storm had occurred while the cat was down there?—but was just one of many risky behaviors Loyd observed. All told, she documented eleven cats entering storm drains (nineteen total events), eleven cats entering crawl spaces under houses (just imagine what might be lurking there), ten cats climbing trees or going onto roofs, and fourteen cats eating or drinking who knows what (in one case, we do know what: someone put a pile of Chex cereal on a tree stump, perhaps for squirrels, but gobbled down by Tigger). Fourteen felines had a total of twenty-eight encounters with cats not from their own household, most of which were friendly, but two of which ended in hissing and growling, though without physical contact. A cat that crawled into a car engine and the opossum incident round out the list.

The significance of these dangerous behaviors paled in comparison to road crossing, known to be one of highest causes of death in outdoor cats, particularly in younger cats. Half of the cats in the study crossed more than five roads during the study; one cat crossed roads twenty-four times (remember, each cat wore a camera for only about a week). This finding agrees closely with the data from the Cat Tracker study.

Overall, almost all the cats engaged in at least one risky behavior during the course of a week's observation, and the average cat was involved in six. Loyd's research article on the study concluded, none too subtly, "Most owners who have found their cat injured or dead from a vehicular collision have subsequently kept their cats indoors; yet, owners whose cat disappears for unknown reasons may be more likely to assume their cat 'adopted' another family, rather than imagine their cat was injured and never found." In other words, cats do dumb stuff, and bad things happen—keep your cats inside for their own good!

The videos also caught four of the cats cheating on their owners and spending time in another house. "They held open the door for him, and he walked in. He just hung out in the house," said the shocked owner of a striped tabby. These two-timing cats were "receiving snuggles and food," Loyd observed; more bizarrely, in one case the camera recorded the homeowner holding a phone up to a neighbor's cat, as if the call were for him.

This feline infidelity, also discovered in Shamley Green and the North Carolina tracking study, didn't surprise me, because a few years ago, I learned of our own Winston's wandering ways. Melissa used to host a weekly gathering in which people from all over the neighborhood would come over for a yoga session. Sometimes neighbors would bring friends that we didn't know. One morning, in the middle of downward dog, Winston ambled through the room, stopping periodically for pets and nuzzles.

To Melissa's surprise, an unfamiliar newbie exclaimed, "Winston, what are you doing here?" To which Melissa replied, "Winston lives here. How do you know him?" Our new acquaintance turned out to live nearby, on the other side of the small creek bordering our backyard. And, we learned, Winston was a regular visitor, entering through their cat door, eating their cats' food, and lying around enjoying the attention of their children. Apparently, he had provided great solace when one of their own cats had died. Checking to see whether he was a stray, they had taken him to a vet, who had located the identification chip implanted near his shoulder and told them Winston's name, though apparently not his address. We now know the neighbors well, and they periodically report on his visits.

The most important finding from Loyd's work concerned the predatory behavior of her subjects. Everyone knows that outdoor cats go after birds, rodents, and other small critters. But how successful are they, and what impact do they have on prey populations? And has domestication affected their predatory prowess?

You would think the way to address these questions would be to watch cats and directly observe them on the prowl. But as we've already discussed, that's very hard to do. In fact, prior to Loyd's study, only one previous project had directly recorded what pet cats catch—Kays's Albany study. In addition to tracking cats, Kays and his team had also periodically watched the cats with binoculars after they'd been located, logging nearly two hundred hours of observation.

In lieu of direct viewing, the way scientists generally have estimated the hunting success of household cats is by recording what they bring home, usually by asking their owners to keep a running tally. There's an obvious problem with this approach: what if the cat kills a prey item and eats it on the spot? Loyd's kittycams were uniquely poised to find out how often that happened.

The videos provided plenty of evidence of hunting talent: a cat pouncing on a leopard frog, another playing with a barely alive lizard;* other cats pummeling a bird† and keeping watch on a bird feeder. Overall, twenty-four of Loyd's subjects stalked prey a total of sixty-nine times; slightly more than half the hunts were successful.‡ Sometimes, the identity of the prey item was hard to discern; in most cases, however, when the prey was hanging from the cat's mouth, right in front of the camera, identification was easy. The sixteen cats that hunted successfully averaged two and a half kills per week;§ Bean was the record holder, bagging two frogs, two rodents, and a lizard.

Loyd's study yielded two big surprises. First, most prey were not brought home. A whopping forty-nine percent of the prey were killed and left in the field, while another twenty-eight percent were consumed on the spot. Fewer than one in four prey were taken home to be eaten there or deposited as a gift to the homeowner. The implication of this finding was profound: previous researchers had been massively underestimating the impact of housecats by counting cat-delivered carcasses. Kays a decade earlier in Albany had come to very similar conclusions, finding a three-to-one ratio of killed prey to presents brought home.

---

*A skink, to those herpetologically inclined.

†An eastern phoebe.

‡By contrast, in Kays's study, only one hunt in four was successful, and in half of those cases, the captured rodent subsequently escaped! The difference may be due to variation in the availability of prey types—for example, there are few reptiles to hunt in upstate New York.

§Although this only refers to the sixteen successful cats. Averaged over all fifty cats in the study, the figure would be somewhat less than one kill per cat per week, and that might even be an overestimate of Athenian cats in general because people who volunteered their cats for the study may have tended to be those whose cats are active outdoors a lot, whereas people with sluggish cats may not have thought their pusses appropriate for the study.

What the cats were hunting was the second surprise. Most public attention has been on the birds killed by housecats. Yet the most common prey item was reptiles, particularly—I'm sad to say—my beloved *Anolis* lizards, specifically the beautiful green Carolina anole. Rodents (chipmunks and voles, to be precise) came in second. Feathered prey, however, were not ignored. Quite the contrary, ten cats watched birds at feeders or birdbaths, and five birds were captured and killed.

KITTYCAMS PLUS CATS BEHAVING badly equals media sensation: Loyd's study was widely reported in print, on the air, and online.

Given all the publicity surrounding cat filmmaking, I would have expected a huge surge in research, with many other scientists following Loyd's lead. Surprisingly, this has not been the case. To my knowledge, only three other research groups have enlisted pet cats as autobiographical videographers.

I don't know why more scientists haven't started projects outfitting pet cats with video cameras, but one researcher said that she was dissuaded from doing so because of privacy concerns. In other words, the fear that cats would inadvertently film people in intimate or personal moments, or that people would worry about that possibility and threaten legal action.

Indeed, for just this reason, Loyd and the Nat Geo team required cat owners to sign an affidavit that they were aware their cat would be wandering around their home taking videos (so be careful when you get out of the shower). Abernathy says that when viewing the videos, if they had discovered anything untoward, they would have immediately deleted the footage, but this nuclear option was never required because the cats' videos were all PG-rated.

Although limited in scope, the three more recent studies—two in

New Zealand and one in South Africa—suggest a basic similarity in pet cat behavior around the world. Cats were mostly inactive, spending about ten percent of their time moving and almost as much time grooming. Almost all the cats in all three studies hunted, and success rates were less than fifty percent; the majority of prey—mostly lizards, insects, and other creepy-crawlies—was not brought back home. Cats seemed to revel in behaving in risky ways: crossing roads, climbing underneath cars and onto roofs, entering storm drains and crawl spaces, and eating and drinking whatever they came upon. Encounters with other cats occurred regularly, sometimes leading to fights, and occasionally cats entered other homes.

A fabulous online greatest hits montage from the Cape Town study illustrates these points, showing cats jumping onto a roof to chase a squirrel, yowling at other cats, watching cavorting dogs, observing little boys both standing on their heads and hiding around corners, staring at birds, pawing at lizards, and carrying live geckos and rodents in their mouths. The highlight, though, was watching an African crested porcupine* amble across the screen and out of sight, the cat wisely choosing not to pursue the prickly rodent.

BETWEEN TRACKING AND KITTYCAM STUDIES, what housecats do when they go outside is no longer such a mystery. Still, there's much we don't know, and more research is needed. Almost all tracking studies, for example, have been conducted in economically well-off Western countries. Do cats behave differently in the tropics, or in China and Thailand? How does their home environment—rich versus poor,

*Several times the size of the American porcupine with quills a foot and a half long.

urban versus rural, living with dogs versus mercifully dogless—affect where they go and what they do?

One particular question concerns differences among cats. I summarized what the average cat does, but not all cats are average. Some range much more widely than most. Some are superpredators. Some cats prey mostly on birds, others on chipmunks. And some are fighters, others are lovers. How do we explain this variation? Right now, we don't know. The good news, though, is that the technology is now available; it's just a matter of the next generation of aspiring young scientists getting out there and collecting the data.

Speaking of variation in behavior, there's one enormous reason that some cats behave differently than others. Some cats don't live in homes with humans catering to their every whim and fancy.

Eighteen

The Secret Life of Unowned Cats

Tens, maybe hundreds, of millions of the world's cats live outside full-time.* Although some are fed by humans and even get some veterinary care from concerned felinitarians, others are completely on their own and have reverted to pre-domestication ways.

This ancestral lifestyle is likely very different from that of household pets. We've already seen that feral cats have larger—sometimes tremendously larger—home ranges. They also are probably much more active, although data are scarce: according to motion sensors

*There are no good estimates of the global population of cats, though the best guess—and it is that—is around 600 million. How many of these are unowned outdoor cats is unknown. In the United States, there are probably somewhere between fifty and one hundred million pet cats and another thirty million unowned cats.

placed on the collars of cats in Illinois, ferals run, hunt, and play more and sleep less than their home-dwelling cousins.

The higher level of activity and movement of feral cats is not surprising—they don't have a bowl of food awaiting them in a cozy kitchen. However, the location-tracking and motion-sensor data paint a very incomplete picture of what these cats get up to during their lives. Unfortunately, observing the behavior of feral cats is even more difficult than watching pet moggies. Imagine what it's like trying to follow around a shy, secretive, downright wild cat. This is exactly the reason the National Geographic Society created Crittercams—to be able to get a glimpse into the lives of hard-to-observe species.

There's one big catch to using Crittercams on such animals. Consider my project with Winston. Every day when I wanted to get some footage, I'd coax Winnie over to me with a treat or some chin rubbing and then quickly place the collar—camera attached—around his neck, snapping the clasp closed in one fluid movement (well, that was my goal; pinning him to the ground with my elbow while I fumbled with the clasp was more usual). Then, when Winston returned inside several hours later, I'd detach the collar, usually while he was munching on kibbles, and remove the camera so I could plug it into my computer to see the day's videos.

Working with feral cats is not so easy. First, you've got to get the camera on the cat. It's going to take more than Friskies Party Mix to attract said puss, and certainly more than an elbow to restrain him. But that's only half the battle—to get your video footage, you've got to retrieve the camera.*

*By contrast, even if you never get a tracking transmitter back, you won't have lost the data because it was already broadcast and received. The transmitters are also cheaper than kittycams, so the financial hit on your research funds is less if the unit is never recovered.

National Geographic has devised workarounds for part two. Some collars are equipped with automated release mechanisms that cause them to drop off the animal at a prearranged time. Recently, they've developed cameras that stream live video back to the researcher. Both of these options, however, add weight to the collar as well as expense, limiting their use to larger animals and bigger research budgets.

Absent these options, there's only one recourse: recapturing the cat and manually removing the collar and camera. And if you think it's hard catching a feral cat once, imagine the difficulty of finding and capturing her a second time!

Hugh McGregor is a problem solver, the sort of person who finds a way to get the results he needs. And the native Tasmanian was not about to let wary felines stand between him and answers to Australia's pressing conservation problems.

The background to his study was the decline in small mammal populations in northern Australia. Cats were the prime suspects, but the case wasn't ironclad because cats had been in Australia for two centuries, yet the declines had occurred only in the last few decades.* One hypothesis suggested that cats had become a much bigger problem recently because increases in the frequency of fires and more intense cattle grazing were eliminating vegetation in open areas, reducing the hiding places prey species need to elude their predators. McGregor's goal was to find out if the wild felines were using some

*Cats were first introduced to Australia in the late 1700s. Within a century, they spread across the entire continent and now occur in all types of habitat, from deserts to rainforests to temperate mountaintops. Australia's two million feral cats have become a major conservation problem because of predation on the local birds, mammals, and reptiles, which have no evolutionary experience with such a talented predator.

habitats more than others, and whether their hunting proficiency was affected by vegetative cover.

The first issue was how to catch the cats. McGregor needed a method that would work multiple times. A trap might work to get a camera on a cat, but as they say, once trapped, twice shy. Some other means would be required to recapture the cat and get the camera back.

McGregor had a different idea than traps: sniffer dogs! Enter Sally, a gorgeous brown-and-white Springer Spaniel with shaggy auburn ears, and Brangul, a barrel-chested black-and-brown Catahoula Leopard Dog, a breed that legend has it was developed in Louisiana to track boars.*

Acquired as eight-week-old pups, the dogs were slowly trained to follow the scent of feral cats. Using the principle of positive reinforcement, they were first taught to recognize the smell of feral cats by presenting them with cat skins and scats.† Then, cat skins were dragged through fields, and the dogs were rewarded for following the trails, which became longer and more complicated through time. Finally, a cat in a cage was put at the end of the trail to teach the dog to associate the smell with the living animal. At this point, at nine months of age, the dogs were ready to get to work.

McGregor's approach to finding cats was to drive around at night in a pickup truck with several people standing in the back scanning the landscape with high-powered flashlights. Cats, like many nocturnal animals, have eyes that brightly reflect light, producing a glowing

---

*Employing dogs to help capture felines was not new—scientists have been using dogs to track down tree mountain lions for decades, for example—but I am not aware of them previously being used to study feral domestic cats. McGregor credits an excellent conservation organization, the Australian Wildlife Conservancy, for suggesting the idea.

†For those unfamiliar with the term "scat," substitute "turd," "poop," or more politely, "dropping."

yellowish-green effect.* Consequently, an effective way to locate nocturnal animals is to use a strong flashlight held near your own head so that the reflected light will rebound to your eyes.

I've done this myself many times, out on a lake or river looking for crocodiles, on safari in Africa looking for lions and leopards, and in the Central American rainforest looking into trees for woolly possums and kinkajous. All of a sudden, you'll see two glowing circles, sometimes red, sometimes green, sometimes silver. Usually you can't make out anything else; it's too far away and too dark to see more than two glowing lights. What is it?

After a while, you get to learn the differences among species. Color and size often can be helpful. For example, the forest is often alive with little silver reflections; size alone tells you they're spiders. Water droplets can be deceptive, but you quickly learn that if you see only one reflected circle, it's probably $H_2 0$. At least in Africa, blocky horizontal reflections, instead of circular ones, are usually from herbivores like antelopes.

Still, many creatures have similar eyeshine—round and green are particularly common—so sometimes you have to get close enough to make out the animal itself. I often take binoculars with me, big ones that are very good in low light. It's hard to hold binos to your eyes while simultaneously aiming a big flashlight at the glowing orbs, which is why I often wear a powerful light strapped to my forehead;

*Many animals (though not humans) have a reflective layer called the "tapetum lucidum" (Latin for "shining layer") at the back of the eye behind the retina. Light bounces off it and passes forward through the retinal cells, giving them a second chance to be illuminated. This double enlightenment is one of the reasons cats have such good night vision, said to be anywhere from three to eight times more acute than humans (the other reason is that their eyes have many more rods—photoreceptor cells capable of operating at low light levels—than we do).

thanks to the night mountain-biking craze,* many different headlamp models are now available. This approach has its own drawbacks, however, notably the moths constantly flying around your eyes and into your mouth and nose.

Moths probably weren't an issue, however, as the headlamped researchers drove around in a pickup at fifteen miles per hour, scanning the surroundings for two large, glowing, yellow-green lights. False alarms occurred regularly. The eyes of dingoes (the wild dog of Australia) look very similar to those of cats, but dingoes are taller and walk away from the light, whereas cats usually sit and stare back.

Another cause of mistaken identity was more surprising. Almost all birds are diurnal, but the nightjar is an oddball. Flying at night requires large eyes that produce a disproportionately large reflection for such a small creature, and the brown chicken-size fowl has a penchant for sitting in the middle of roads, its eyeshine visible far away to oncoming cars. Still, the glow is not as bright as a cat's, and only toward the end of a long night would fatigued spotters be fooled.

Once a set of peepers was identified as feline, the researchers would yell to the driver, the pickup would screech to a halt, and out would go the dogs. Muzzled to avoid harming the cats, Sally and Brangul would start sniffing around until they picked up the scent; then, game on! The dogs would take off, sometimes at high speed, handlers frantically trying to keep up.

Sally was the more enthusiastic of the two, sometimes a bit too eager. "Sally was full of beans and enthusiasm, but she wasn't the smartest in terms of cat hunting, and there were many instances where the cats outwitted her," McGregor said. "It was that classic

*I use this term advisedly, as few hobbies seem more lunatic than zooming down a rough trail in the dark.

cartoon image of the dog chasing the cat, the cat stopping still, and the dog jumping right over the cat and continuing to run."

Brangul, by contrast, was less gung ho, but savvier. The two made a good team, and by the end of the study, they never missed a cat in open habitat; even in more difficult, cluttered terrain, success rates were over fifty percent. The length of a tracking session was anywhere from thirty seconds to sixty minutes, depending on how much of a head start the cat had and how far she ran before ducking into a hole, climbing a tree, or being cornered by the dog. At that point, Sally and Brangul's work was done and it was up to McGregor to collect the cat.

Easier said than done. In many cases, when the cat was on the ground or up a small tree, she could be snagged in a net and then placed in a canvas holding bag. Sometimes the cat could even be pulled out of a hole by her hindlegs. Amazingly, despite handling more than three hundred cats in his career, McGregor has only been badly bitten once. "Friends I know studying parrots have it far worse; cats are a walk in the park in comparison," he says modestly, but those of you who've given your cat a bath realize how skilled McGregor must be.

Oftentimes, however, the cat had climbed too high in a tree for the researchers to follow. In that case, McGregor would shoot the cat in the upper thigh with a tranquilizer dart; he and the other researchers then would wait underneath the tree holding a sheet between them, fireman-style. Eventually, the groggy cat would topple out of the tree to a soft landing. They didn't miss a single cat.

THIS IS A GOOD POINT to discuss animal welfare considerations in scientific research. These procedures sound unpleasant. Being chased by yapping hound dogs must be traumatic, to say nothing of being shot with a tranquilizer dart.

Indeed, just about any behavior research that does more than observe an animal from a great distance leads to the possibility of bothering the study subject in some way. Even wearing a kitty camera may be uncomfortable at first.* McGregor's cameras weighed about three percent of the cat's body mass. That's equivalent to walking around all day with a very large pair of binoculars around your neck.† The cats didn't sign up for this!

The topic of animal rights vis-à-vis scientific research has been vociferously debated for more than half a century. At the extremes, one side argues that any research that inflicts even the mildest annoyance on an animal is impermissible, while the other side suggests that the quest for knowledge justifies any research. Relatively few people hold the former view and hopefully even fewer the latter, but where in the middle do we land? Knowledge and animal welfare are apples and oranges; how do we come up with a formula that relates learning gained to discomfort caused?

Entire books have been written on this question. Many issues are

---

*Surprisingly little research has investigated how cats respond to wearing transmitters and cameras. One study found that the weight of radio-transmitter collars affected how far pet cats wandered: the lighter the devices, the farther the cats moved from home. The effect was slight, however: when cats wore devices weighing one percent of their body weight, they traveled on average 137 feet from home, while the same cats only moved away 119 feet when wearing devices three times as heavy. Another study found that some cats bearing commercial trackers (different models than the one I put on Winston) scratched themselves with their hindlegs or shook their heads and bodies more than control cats not wearing the devices (the study recommended that companies developing these trackers should pay more attention to what they call the "wearability" of the devices, with the comfort of the cat, rather than the aesthetic sensibilities of the paying customer, in mind). Based on these findings and observations that the cats in these studies behaved in normal ways—hunting, fighting, cruising about—I conclude that wearing a tracker or camera does affect cats, just not much.

†Though, of course, human and cat necks are built differently, so the sensation may not be equivalent.

pertinent that I won't get into here. For example, does it matter how closely related the animals are to humans? How much pain can they feel (to the extent that we can know)? And what about body size, cuteness, and charisma? Many more people are concerned about research on cats and horses than on mice and snakes, much less cockroaches. Is that ethically justifiable?

My view is that human history has been a journey of discovery about ourselves and the world around us. Knowledge for its own sake is intrinsically valuable, and it is also useful in building a better world.

But we mustn't forget that animals are living, breathing organisms. They have some degree of self-awareness, and most can feel pain. To me, imposing some minor degree of discomfort is acceptable in the quest of knowledge. Placing a collar on a cat that causes a bit of itchiness for a while doesn't seem that onerous.

But causing greater levels of discomfort, including pain and the possibility of death, requires greater justification in terms of the knowledge to be gained. Moreover, if the research is justified, all reasonable steps must be taken to find alternatives or minimize the distress and pain. These issues are considered by the review committees that must approve all animal research at universities, government agencies, zoos, and similar institutions. Of course, this is the apples and oranges aspect—people will differ on where the balance should lie. Some would consider subjecting a cat to an annoying collar just to learn where he goes to be unacceptable. The best approach is to weigh each endeavor in its context—what is to be gained, and at what cost to the study subject?

With regard to McGregor's work, not only are we curious about the private lives of cats, but we also need this information to help understand cats' impact on the environment, and if it's detrimental, to devise solutions. Because catching the cats with traps is difficult, especially after they've been trapped once, no practical alternatives exist to capture these wary, wild felines. Moreover, McGregor's team worked very hard to ensure the safety and minimize the discomfort of the cats. None of them sustained an injury more severe than a bruise where the dart hit. To me, the knowledge gained is worth scaring and mildly hurting some cats. Not everyone will agree.

Now back to Brangul and Sally.

ONCE THE CATS were in custody, they were quickly outfitted with their kitty camera necklaces. These were not the Nat Geo Crittercams—the project began before Loyd had published her work. For years, McGregor had talked to engineers and others trying to get someone to build him a working kittycam, but to no satisfaction. So he decided to take matters into his own hands.

His first attempt was simple: a GoPro camera suspended from a cat collar, only capable of recording for two hours* and only during daytime. The result? "The footage I got back was incredible! Hunting lizards and quails. Jumping around rocks. I was even able to see on camera where it did a poo, then go out to that location and collect it for a sample."

McGregor was hooked. It didn't matter that the subsequent six attempts were complete failures: two hours of footage shot while two cats slept, a camera malfunction, a cat active when it was too dark to make anything out, another camera failure, then another snoozing feline.

*The high-resolution of GoPros is great for producing fabulous videos (much higher quality than Nat Geo's kittycams), but they come at the cost of draining the batteries much more quickly.

McGregor persevered. He learned how to modify the cameras, making them tougher, giving them night filming capability, and adding a motion sensor so that they wouldn't record when the cat was inactive. Even with all that, only one deployment in every three yielded usable videos.

But what videos they were! A mom cat wrestling with her kitten! A cat climbing a hill to watch a sunrise!

McGregor had already placed GPS units on thirty-two cats to track their movements. The data beamed back from the GPS units allowed McGregor to investigate whether the cats were preferentially using open areas. And that's exactly what they were doing, strongly favoring habitat patches where the grass was short or where there had recently been an intense fire, presumably because in those areas, prey had few places to hide.

It's always gratifying to collect scientific data and discover that the results are just as you predicted. But what's really exciting is when you learn something you didn't know, even better if it's something you'd never even considered a possibility. And that's what happened when McGregor started looking at the GPS location data.

By definition, a home range is the area an animal uses during the course of its normal activities. McGregor's cats were generally typical in their roaming, using large areas like other feral cats, averaging two thousand acres for males and half that for females. But seven cats did something never seen before in any *Felis catus*. All of a sudden, one day they took off, moving fast and in a straight line, leaving their home range and traveling as far as nineteen miles to another area (each cat did this on different days and to different areas). Once there, most of them wandered around a circumscribed area—a home-away-from-home range—before returning to their real home range on average fifteen days later.

Why would a cat leave the area it knows well and make a beeline—so straight in one case that McGregor initially thought the GPS unit was malfunctioning—to a faraway place that she had probably never previously visited? The answer is simple: the cats were traveling to areas that had recently experienced intense fires. Most of the fires had occurred within the preceding two months—the quickest response was five days post-inferno—and the more recent the conflagration, the longer the cats stayed before returning home.

Detailed examination revealed that this behavior was specific to intense fires (defined as fires in which all trees were scorched and no ground cover remained unburnt). McGregor scoured the data and found that of the cats for which an intense fire occurred within eight miles of their home range, almost all visited the site. By contrast, none of the twenty-two cats whose range was within eight miles of a mild fire went to investigate.

These amazing journeys raise two questions. First, why did the cats take trips to locations far from their normal haunts? The answer: opportunity. In areas with massive fires, most of the vegetation had been burned to the ground—picture a flat, blackened landscape with piles of ash and charred sticks poking up here and there. For the local mice and other small critters that survived the conflagration—and many animals are quite adept at finding refuge and surviving even major fires—there aren't many places to hide. To a cat, it's a snack bar with the candy lying out in the open, easy pickings.

Of course, testing this hypothesis is difficult. McGregor wasn't even aware of these journeys until after they had occurred, so directly observing hunting was not possible. However, another study by McGregor and colleagues provided supporting evidence. The researchers burned one area, keeping a second area intact as a control, and monitored rodent survival. The burn itself had no effect on mortality, but subsequently, rodents in the burnt area died off at a much higher rate as a result of predation.

The second question is harder to answer. The cats headed straight for the burned areas, sometimes several months after the fire occurred. How did they know an intense fire had occurred in an area so far away, and how did they find their way to it so long afterward? McGregor ticked off the possibilities: the cats see the red glow of the fire or smoke on the horizon, they smell the smoke or ash, or they observe other animals moving to the fire (birds of prey are famous for doing so to catch fleeing prey).

All of these are possible, but they all have problems. If the cats saw the fire or watched other species, they'd have to remember the location for days or months. Alternatively, instead of relying on memory, the cats may have been able to smell the fire long after it had gone out. But would the aroma have been intense enough months later for cats to track it to the source? And can they really tell the difference between the lingering smell of an intense fire, to which they respond, and that of a milder one, which they ignore?

We don't know, but McGregor tracked one cat walking to the top of a hill, staying there for several hours, and then walking directly to a fire site about five miles away. Could he see the fire scar? Or was he orienting to the smell? Who knows? McGregor suggested that this surveying behavior might be the key, noting that videos show the cats spending "a lot of time looking and watching the horizon. It's almost a bit like *The Lion King* at times."

McGregor's kitty-camera studies complemented what he learned from the GPS units. Over the course of three years, thirteen cats wore cameras, sometimes more than once, for a total of twenty-three sessions producing eighty-nine hours of usable video. The cats were active forty-seven percent of the time, very similar to feral cats in Illinois farmland and substantially more than pet cats in Illinois or Georgia.

Given the relatively limited filming period—averaging less than four hours per session—you might wonder whether any predatory behavior would be recorded at all, especially because the videos were recorded shortly after cats had experienced the trauma of being chased by hounds, caught, and handled by researchers.

Ye of little faith! Twenty-one of the twenty-three videos featured hunting events, one hundred one in total, thirty-two of which were successful. The most popular item was frogs, constituting nearly half of all prey, followed by six rodents, three lizards, two snakes, two quail, a nest of bird eggs, and a locust.

Actually, "popular" may not be the best word to describe the frogs, because half were not eaten. Many Australian frogs produce noxious skin secretions as a defense mechanism, so the cats may have regretted their choice of prey. Indeed, in one of the videos, a cat is seen catching a frog, then dropping it and making the sort of exaggerated mouth movements that both cats and children make when they've tasted something unpleasant and are trying to remove the sensation from their tongues.

The major focus of the study was the cats' hunting success. The GPS tracking data already had shown that the cats prefer open areas. Was that because the hunting was better there? The answer was a resounding "yes!" Cats were four times more likely to be successful when they attacked prey in open habitats (seventy percent success rate) than when in dense grass or complex rock piles (seventeen percent successful).

Some of the videos illustrate the difference. In one, a cat is walking in an open area. Suddenly, a rock rat runs across from the left. A blur of activity. I can't tell what's happening; maybe the cat's pinning the rodent with its paw (just as in Loyd's videos, I found a lot of the action to be disorienting and hard to decipher as the point of view tilted and bobbed chaotically). Regardless, in a moment, the rodent is in the cat's mouth.

Contrast that success with a cat moving through dense grass; suddenly, a flurry as something—four viewings later, I decide it must be a bird—erupts from the grass; the cat's head swivels up and watches it fly away. In several other videos, cats clearly are peering into dense vegetation. There must be something there—the cat pushes its head through the grass one way, then the other—but we never see the quarry, and the cat goes away hungry.

Not all potential prey are able to hide within thick vegetation, however. Some are just too large. One video shows a cat grabbing a sizable snake—too big to get lost amidst the jumble of grass blades—and pulling it out of the tussock. And then a big surprise. That's not just any snake. It's a western brown snake, one of Australia's many highly venomous snakes, a lethal threat to humans and cats alike! No matter. The cat rapidly dispatches it and then spends ten minutes chewing off its head before eating the remainder of the snake.

It turns out that's not standard snake procedure, however. Another video shows a cat catching a much less venomous shovel-nosed snake. No special treatment, no decapitation. Rather, straight down the hatch, the tail still wriggling as it disappears into the cat's gullet. Drawing grand conclusions from two data points is always dicey, but it seems that somehow, whether through evolution or learning, *Felis catus* has gained the ability to distinguish dangerous from non-dangerous prey and react accordingly. Not bad for a species that has only been in Australia for a couple of centuries!

Overall, the thirty-two prey items the Outback cats caught in eighty-nine hours average to 0.36 prey per hour, or 8.2 per day. Admittedly, the prey were small, but that's still a lot! Remember that, by contrast, the cats in Athens, which were hunting similar-size critters, were only catching two and a half prey per week. That's a twenty-five-fold difference!*

Of course, the explanation is obvious, the same reason that pet cats have smaller home ranges and are less active: they've got a dish full of kibble waiting for them on the other side of the cat flap. No need to overly exert oneself except for the occasional appetizer.

BUT MAYBE IT'S THE cat flap and not the plentiful kibble. Perhaps feral cats hunt more not because they're hungry, but because that's the outdoor cat way of life: put a cat outside and maybe her instincts kick in and she starts hunting regardless of how full her belly is.

In theory, distinguishing between these possibilities should be easy: keep cats outside all the time, but feed them amply. And in fact, that experiment is done all over the world in places where people provision unowned outdoor cats.

If such outdoor colony cats hunt because it's their nature to hunt, then colony cats should behave like those in the Outback. But if predation is driven by hunger, then their predatory behavior should be more like that of well-fed pet cats.

Nowhere have colony cats been studied more thoroughly than on Georgia's Jekyll Island. Once an enclave for some of the world's wealthiest, Jekyll is now a state park, home to scads of summer tourists, abundant wildlife . . . and nine cat colonies. And who better to study the behavior of these cats than Sonia Hernandez and her kittycams team?

---

*And that's not even accounting for the majority of the Georgian cats that didn't hunt at all. If we include all fifty-five cats in Loyd's study, then the catch per cat was fewer than one prey item per week, less than two percent of the Outback cats' toll.

A trap-neuter-return (TNR) program was organized by a local homeowner distressed by the multitude of stray cats, including, in his words, some of the "meanest and unhealthiest cats we had ever encountered." After a program of removing or treating sick cats, putting some up for adoption, and neutering and returning others to their point of capture, the feral cat population was drastically reduced. Feeding stations were set up to keep the cats healthy and, I'm guessing, to try to minimize their impact on the native wildlife (though the colony managers touted the cats' role in "controlling the overpopulation of . . . snakes" as a reason to keep them around).

Hernandez was aware of the Jekyll Island colonies living around the feeding stations and thought they would be suitable subjects for the next round of cat-borne video research. Her request to the Jekyll Island Authority for permission to conduct the studies was swiftly granted: the director of conservation, who had completed his graduate work in Hernandez's department, also wanted to know more about the cats.

The goal of KittyCams 2.0, as Hernandez christened it, was to investigate how TNR colony cats differ from pets in their behavior and environmental impact. A novel second goal was to see how cats living outside all the time interact behaviorally with the abundant and diverse wildlife on the island.

Logistically, working with colony cats was easier than dealing with ferals. No need for sniffer dogs to track down the cats—they showed up on cue at the feeding stations every morning when food was put out. But still, these are cats that mostly grew up in the wild—you couldn't just walk up and put a collar on them. They had become accustomed to the two colony managers, but the presence of anyone else would send them dashing away. For this reason, the original plan had been for the colony managers to be the ones to put the camera collars on the cats. But for various reasons, that proved unworkable. Instead, to run the project Hernandez hired Alexandra Newton McNeal, who had just graduated from UGA with bachelor's degrees in both wildlife biology and fisheries science.

Predictably, the cats wanted nothing to do with McNeal. For six months, she accompanied the caretaker on her daily rounds, letting the cats become accustomed to her presence. Eventually, she was able to walk right up and give them a friendly back scratch as she sprinkled kibble on the ground.

The time to deploy the cameras had arrived. With a lightning-quick move, she'd grab the cat by the scruff of the neck with one hand and quickly slip an expandable collar over her head with the other—the kittycams were in place before the cat knew what was happening, and McNeal was never bitten.* Some would run away for a couple of minutes before returning to finish their meal; others were unfazed and immediately resumed chowing down.

For those cats more resistant to McNeal's wiles, she had one last trick up her sleeve, one familiar to most cat owners: break out the wet food! Just like every cat I've ever known, the Jekyll strays quickly learned to respond to a particular metallic screech—"They heard the sound of a can opening and they'd come sprinting out of the woods," McNeal recalls. Only the most wary of felines were able to resist the allure of Fancy Feast.

Getting the cameras back after their twenty-four-hour deployment

---

*To this day, Nat Geo's Abernathy is still amazed by McNeal's dexterity. When he was asked about the Jekyll Island project, her technique was the first thing he mentioned. McNeal had been prepared for the worst and had developed a protective armory including special gloves and a shield made from a milk jug, none of which proved necessary. By the end of the project, she was more worried about ticks and chiggers than injuries from the cats.

could be more challenging. Sometimes, McNeal said, the cats would remember that "hey, yesterday this lady grabbed me and put a camera on me. I don't want to get anywhere near her." But eventually they'd allow her to sidle up and retrieve the device.

Sometimes, though, a cat would return the next day to the feeding station cameraless. Like pet owners, all cat researchers use breakaway collars; if the cat gets caught on something, better to have the collar fall off than risk harming the cat. For just that reason, the cameras were equipped with radio transmitters. Usually when the cat showed up sans neckwear, McNeal had no trouble locating the camera by breaking out the antenna and using old-school radio-tracking technique.

In one case, however, relocating the camera was a little trickier. The cat in question was fed at one of the more suburban feeding stations, located near a residential area. McNeal felt a little self-conscious walking through the neighborhood "with a giant antenna," especially when the beep-beep-beeping was "going off like crazy" as she approached one house. Initially she thought the cat must have lost the collar in the front yard, but the signal led her straight to the house. She knocked on the door, but no one was home.

What to do? It seemed like the signal was coming from within the house. But as she stood at the front door, idly swinging the antenna while she decided what to do, she noticed that the signal seemed slightly louder—if that was possible—to the right, in the direction of the carport. And in front of the auto bay—perhaps it was trash collection day?—was a garbage can.

Dumpster diving wasn't part of her job description. But curiosity—or maybe it was determination—got the better of her. She opened the lid, ready to rummage. No need. There it was, right at the top, the intact camera and lots of tiny bits of collar snipped into pieces.

"Man, you could have just called the number [printed on the

camera] and I would have come and picked it up," she thought. But she didn't stick around to learn more. Some locals were not so keen on the project, fearing that the scientists' intent was to eventually remove all the cats. Or perhaps the cat, despite feasting regularly at the feeding station and having no sign of ownership, was actually a pet. No matter. Mission accomplished, camera recovered.

Collar casualties notwithstanding, the project was a great success. Twenty-nine cats wore cameras for an average of twenty-two hours each—nearly seven hundred hours of footage. The job of viewing the videos fell to McNeal, but like Loyd before her, she was upbeat: "At times spending hundreds of hours watching cats grooming themselves would make you go crazy. But a predation event or some cool interaction made it all worthwhile."

Just like the Athens pet cats in Team KittyCams' first study, the colony cats were layabouts, spending only ten percent of their time moving around, hunting, and feeding. Home range was not explicitly calculated, but cats only rarely changed feeding stations, suggesting that their home ranges were small, like those of pet cats. Score one for the hypothesis that provisioned cats, even if they live full-time outside, will be as slothful as indoor-outdoor pets.

It was thus a surprise to discover that the cats were hunting much more than the cats of Athens. Remember that in KittyCams 1.0, fewer than half the cats exhibited hunting behavior; on Jekyll, the proportion of hunters was nearly twice as high.

And not only were they hunting, they were succeeding—the island cats averaged more than six kills per day, approaching the hunting prowess of the Outback cats. But unlike in other studies, nearly half of their prey were invertebrates, mostly crickets, grasshoppers, and cicadas, but also beetles, moths, dragonflies, and spiders. By contrast, only twenty-one percent of the Athens prey and three percent in the

Outback were invertebrates. Among the vertebrates taken on Jekyll, frogs and lizards made up the vast majority, but various small mammals were captured, including a squirrel, a rabbit, and a bat.* And as for those claims of snake control? Although two cats spent three hours stalking and harassing a three-foot-long snake,† none were caught. And one last point of interest: the colony cats had a greater than fifty percent capture rate for all types of prey save one—only two of twelve birds were successfully stalked.

The high kill count might seem to support the idea that even if cats are well fed, they'll hunt—it's what cats do. In support of that hypothesis, at the one feeding station where food was available throughout the day, located at the colony manager's house, three of the four cats hunted.

Although Hernandez and her coauthors lean toward this explanation in their paper, I'm not so sure. In Athens, the pet cats ate only a quarter of the prey they caught. If the Jekyll cats were hunting just to get their ya-yas out, we might have expected a similar rate of abandonment, but in fact, more than eighty percent of the prey were consumed. My conclusion—a possibility mentioned in Hernandez's paper—is that the cats weren't getting enough to eat at the stations and were supplementing their commercial diet with wild game.

Regardless of whether the Jekyll cats were hunting out of hunger or not, their environmental impact was less than that of the Outback

---

*That's right, a bat. How does a cat catch a bat? Unfortunately, we don't know, as the video starts with the flapping bat in the cat's jaws. Because the cameras didn't record when the cats were inactive, McNeal surmises that the bat flew or dropped right in front of a resting cat. This event, incidentally, was the first that came to McNeal's mind when I asked her about the most memorable videos. Incidentally, the observation is not as surprising as it may seem: cats are regular predators of bats in many places around the world.

†Probably a black racer.

cats. Not only was the average number of prey caught on Jekyll only three quarters of that in the Outback, but the average size of the prey was substantially smaller thanks to all those creepy-crawlies. In other words, the feral, unfed cats of the Outback were catching and eating more and bigger prey than the catered cats of Jekyll, and thus having a bigger effect on local ecosystems.

Jekyll Island is a lovely place, looking just like you'd expect for a popular state park on the coast: beaches, forests, golf courses, tree-lined streets. Plenty of natural areas for hiking and birdwatching. Imagine putting out piles of cat food in a place like this. Think cats would be the only diners to show up? Of course not. Raccoons—as many as a dozen—were usually waiting at most of the feeding stations when McNeal showed up in the morning; at several spots, black vultures would descend as soon as she left. Opossums also occasionally made an appearance.* In some places, the colony managers left out extra food to accommodate all comers.

I've seen Winston and Jane interact with raccoons and opossums in my own backyard. Occasionally, there's some hissing and posturing. One evening, Leo, all ten pounds of him, charged at a much larger raccoon on the other side of a sliding screen door, sending the masked bandit hightailing it off the porch. For the most part, though, they ignore each other, ships passing in the night.

And so it was on Jekyll. During the course of the study, 142 nonpredatory interactions were recorded between the cats and other species. In most cases, they peaceably coexisted, sometimes dining within a foot or two of each other.

---

*I was surprised that more species weren't detected. Where were the gray foxes and armadillos? Luckily for the cats, coyotes don't occur on Jekyll. Not so luckily, the first bobcats in a century turned up just before the study began.

A few interactions with raccoons, however, were less pleasant, involving hissing and batting. Although nothing more serious was observed, Hernandez, wildlife disease expert that she is, pointed out that the frequency of such close interactions increases the possibility of disease transmission from wildlife to cats, and then potentially from cats to caretakers. In addition, transmission can go the other way, from cats to wildlife.

A colony cat and a raccoon at a feeding station.

TECHNOLOGICAL ADVANCES WILL continue to lead the way in shaping how we study free-ranging felines. Not only the ever-better cameras and trackers constantly being developed, but also sensors that can record how fast a cat is moving, how sharply it turns, and who knows what else? Researchers—led by Alan Wilson, who developed the equipment for the Shamley Green project—have already developed GPS collars that can precisely track the speed, acceleration, and turning angle of cheetahs as they sprint after their prey. Just think what Fitbits and Apple Watches can do for us, then downsize them to kitty versions. Not only will we soon be able to learn exactly what cats are doing outside, but some companies are even exploring the potential health benefits of feline wearable devices.

There's much left to learn, but cat trackers and kittycams have already made clear that there's no one-size-fits-all *Felis catus*. Cats living in different circumstances interact with the world in very different ways. Even household cats living in similar circumstances can vary greatly in their outdoor activities, probably as a result of differences in their particular experiences growing up, as well as in their genetics.

Cats that grow up out in nature, away from people, seem to be very similar to their forebears in many ways, such as the size of their home range. I say "seem" because we actually don't know all that much about the North African wildcat and related species. Hopefully the new technology will be applied to these species as well, so we can get a better understanding of how they live their lives and the extent to which feral domestic cats revert to their ancestral way of life.

Nineteen

Good Stewardship, or Keeping Your Race Car in the Garage?

Pet cats that go outside cause problems and get into trouble, whether it's catching birds and chipmunks or crossing big dogs, mean cats, and fast cars. In many places, coyotes and even mountain lions are a real threat. And then there are the invisible risks: getting a disease such as feline leukemia, feline immunodeficiency virus, or bartonellosis, or picking up a nasty parasite (outdoor cats are nearly three time more likely than indoor cats to be infected with parasites).

An oft-quoted statistic is that the average life span of indoor cats is seventeen years, whereas cats that spend most of their time outside live only two to five years. Unfortunately, despite much digging, I have not been able to find the basis for this claim, and the discrepancy seems extreme to me. Still, I would be astonished if indoor cats don't live substantially longer, on average, than their outdoor-traipsing counterparts.

There is an easy solution to this problem: don't let your cats outside. Many conservation organizations have programs encouraging just that. Portland's Cats Safe at Home initiative promotes the construction of catios, enclosed outdoor spaces that give your puss a feel for the outdoors without any of the risks. The American Bird Conservancy has its Cats Indoors initiative; Nature Canada's version is Keep Cats Safe and Save Bird Lives. Similar programs occur worldwide.

And it's not just conservation organizations. The Humane Society of the United States strongly advocates keeping cats indoors, as does People for the Ethical Treatment of Animals and many other animal welfare groups.

These efforts seem to be working. The percentage of cats kept indoors has steadily risen over the last twenty years—now more than two US pet cats out of three are indoors only. But this sentiment is not universal. In the United Kingdom, eighty percent of pet cats are let outside, and in New Zealand, a whopping ninety-two percent venture through backyards, fields, and forests. Do Brits and Kiwis not care about their cats?

Of course, they love their cats just as much as Americans, but they're more concerned about their cats' psychic well-being than their physical welfare. "Ideally all cats would be allowed access to outdoors to express their natural behaviour. . . . Cats have a natural tendency to explore so allowing them access to the outside world gives them mental stimulation and reduces stress," declares Cats Protection, the self-proclaimed "UK's leading cat welfare charity."

This disagreement highlights an aspect of feline domestication we haven't yet broached. Given the state of domestication of *Felis catus*—the extent that they've evolved from their ancestral roots and the degree to which they still maintain their inner wildcat—is it unnatural, perhaps even harmful, to keep them indoors? The Brits would seem to think so.

But perhaps this view is just romantic nostalgia for a wilder, less civilized world of bygone days. Cats may already have forsaken the savanna for the comfort and safety of the sofa. And even if they're still in evolutionary transition, isn't this just a standard way station in the domestication process of any species, the stage when we humans take control and direct the process, counting on selection to evolve the species in the direction we choose?

I'VE GOT MY OWN EXPERIENCE with these issues. As you may recall, the feral mother of Jane and Winston was run over when they were two weeks old. Three and a half months of bottle feeding and kitten antics later, J and W entered our household. Given that they had spent almost their entire lives indoors, we didn't expect to have much trouble keeping them as indoor cats.

It didn't work out that way. Upbringing notwithstanding, they were desperate to get out. Both fast and clever, they constantly outwitted us, zipping past when we opened a door, sneaking out when they detected an opening. Plus, they quickly developed the annoying habit of using the sliding screen doors as scratching posts. We relented.

I am not proud of this. It was a battle of wills, and whether the cats' willpower was particularly strong or ours particularly weak, I'll leave to you to decide. Regardless, they became indoor-outdoor cats. We vowed to do better the next time.

That time arrived nine years later in the form of Nelson, who entered our household as a four-month-old kitten. We intended to honor our pledge and keep him indoors, protecting him from disease, dogs, cars, and thieves (he is pedigreed, after all!), and sparing the local birds and other small animals.

We didn't think that would be much of a problem. Nelson was raised for the first months of his life in the home of a woman who breeds European Burmese cats. He received lavish attention and had lots of toys, many structures to climb and posts to scratch, and other kittens with which to interact. In all this time, he never set foot outside. Not having learned our lesson from Winston and Jane, we figured he wouldn't have any interest in—perhaps even be afraid of—the outdoor wilderness.

We couldn't have been more wrong. Almost from the day we got him, Nelson yearned to go outside. He'd stand by the sliding glass door, staring outside, sometimes plaintively mewing or scratching at the glass. When any door was opened, he'd make a mad dash for freedom. Eventually, he learned how to open the sliding screen door by standing up and placing his front paws on the screen, then shifting his weight to his left. The screen (whose lock was not functional) would obligingly slide to the left, and he'd bolt out into the backyard. We had to keep that door closed for months—no cool breeze for us—until we replaced the lock.*

Despite the screen door caper and otherwise being outsmarted, we've been fairly successful at keeping Nelson from going outside unsupervised. I do have to admit that it makes me sad seeing him staring out the door, so clearly wanting to venture abroad. Sometimes he stands by a door, his plaintive cries expressing his displeasure at not being on the other side of the glass. It hurts my heart to see him so wanting to go outside.

---

*He also once managed to slit the bottom of another screen door. We were perplexed as to how he was getting outside—joking about "a hole in the house"—until I happened to be watching when he ducked through the undetected opening.

As a result, we've tried to give Nelson supervised outdoor experiences. First, we took him for walks on a leash in the backyard. He doesn't seem to mind wearing his harness. Indeed, when I pull it out, he comes running, purring loudly.

But walking a cat is not all it's cracked up to be. In fact, it's downright boring—at least walking Nelson is. He spends most of the time standing still, sniffing and looking around, and when he does move, he heads straight into the bushes where I can't follow and the leash gets tangled.

So we've gone to plan B. I put a cat tracker on him—the same one used to monitor Winston's ramblings through the neighborhood—and put him out in the backyard. I let him wander on his own for an hour or so, but unlike with Winnie, I keep a close eye on him so that he doesn't find a hole in the fence. When he does go over or under the fence, the tracker allows me to return him quickly, including once from the garage of the doctors next door; retrieving him from the mean guy's backyard that abuts ours was more tricky.

Nelson certainly enjoys his adventures outside, but in retrospect, I'm not sure we've taken the best approach. Rather than sating his outdoor desires, I fear we've just stoked them. The next day, sometimes even just a few hours later, he's back at the door, pawing and meowing, seemingly as desperate as ever to get out. Still, I'm pleased that we've kept our resolve and not capitulated to his demands for unrestricted outdoor time.

PROPONENTS LIST MANY ADVANTAGES to cats going outside. Exercise is an obvious one: as we've seen, outdoor cats can cover a lot of ground, but indoor cats don't have this opportunity (though the giant hamster wheels now sold for cats may solve this problem for felines lucky enough to have one). Some evidence suggests that indoor cats are more prone to obesity, which in turn can lead to a wide variety of other ailments like diabetes, heart disease, and mobility issues.

The advantages of going outside extend beyond physical fitness. Outdoor cats get to do what cats have done since time immemorial—hunt, sniff, inspect, patrol, explore. Roll around in the dirt. Climb a tree. Experience the adrenaline rush of a barking dog nearby. In other words, to live cat life to its fullest. As a bonus, they escape stressful conditions inside such as loud noises, annoying children, and unfriendly fellow pets.

Some believe that cats kept indoors will be frustrated by their inability to satisfy their cat desires. This psychic harm, it is claimed, will manifest itself in inappropriate behaviors, such as biting, scratching the furniture, not using the litter box, and developing neuroses. There's very little scientific evidence to support these views—appropriate studies on pet cats for the most part have not been conducted. And certainly there are many cats that seem quite content with an indoor-only existence.

Research on other animals, however, clearly reveals the psychological harm of living in an unstimulating place. Anyone old enough to have visited zoos several decades ago remembers tigers, polar bears, and similar creatures pacing back and forth or engaging in repetitive behaviors nonstop. There can be no doubt that keeping intelligent animals, such as carnivorous mammals, in small enclosures with nothing to do is psychologically destructive and often physically harmful as well.

For this reason, zoos now emphasize the importance of "behavioral enrichment," giving animals new places to explore, new challenges to solve, keeping life interesting and unpredictable. All of that is possible with a mindful approach to the environment itself.

The same applies to keeping cats indoors. All experts—from PhD behavioral scientists to cat whisperers on TV—agree that cats need mental stimulation, such as a variety of toys, boxes to jump in, and high places to ascend. And novelty is important—toys should be rotated, furniture rearranged, food hidden, new smells provided. Cats like to hunt, so playing with them in ways that let them express their predatory behaviors is particularly important.

None of this is news to anyone who owns a cat, and the recommendations are standard fare in cat self-help books.* Nonetheless, few researchers have conducted studies on indoor pet cats to evaluate the effects of enrichment. One exception is a study in Tennessee that showed that cats that are played with more display fewer behavioral problems (though being based on a survey of cat owners rather than a controlled experiment, the study could not determine which was cause and which was effect: are behavior problems the result of lack of play or do owners play less with problem cats?).

So who's right? The Americans, who mostly keep their cats inside, or the members of the British Commonwealth, who mostly don't? If the issue is the life span of the cat or his impact on the environment, then the Americans are definitely right. Alternatively, if the major concern is the psychic well-being of the puss, then it's unclear. We simply don't know whether a loving home with plenty of toys and other stimulants is sufficient compensation for not prowling the great outdoors.

And keep in mind as well that not all cats want to go outside. A feline behavior specialist told me that if given the chance, half the indoor cats she knew would run out the door, then immediately turn

---

*A study in Australia showed that cat owners who felt they were capable of keeping their cat happy inside were less likely to let them outside. As a result, a major component of one Australian program is educating people about ways to improve the lives of indoor cats.

around and run back in. One important consideration is whether a cat has spent his entire life indoors. For obvious reasons, formerly outdoor cats are more likely to be upset if they're confined to quarters, whereas lifelong indoor felines may have no idea what they're missing.

But let's be honest here: sometimes the decision whether to let cats outside is more about us than about them. Who wants to clean a litter box when the whole world could be the cat's outhouse? Not to mention outdoor rodent control services (the world seems to be divided between those who love chipmunks and those who hate them).

Probably a bigger motivation is that for some, part of the joy of having a cat is the spectacle of a wild beast in a backyard Serengeti. As one science writer puts it, "To see a cat outside is to see a creature in its element: stalking through the grass, climbing trees, battling for their territory. The predator, the wild cat, the king of the jungle . . . You could keep your cat indoors and give him a comfortable, safe life. You could also keep your race car in a garage."

IS THERE AN INTERMEDIATE COURSE, a way to let cats roam without harming wildlife? Ideas come in two flavors: make cats more detectable by potential prey or reduce the desire of cats to hunt.

One old idea—enshrined in an ancient fable sometimes falsely attributed to Aesop—is to "bell the cat" so that every time he moves, the jingle-jangle will alert potential prey. Seems like a great idea, right? Only one problem. Cats are smart. They learn to move in a way that doesn't make the bell jingle.

A new twist is to alert potential prey visually, rather than aurally, by dressing a cat in a garish collar, making it hard for Oliver to launch a surprise attack.

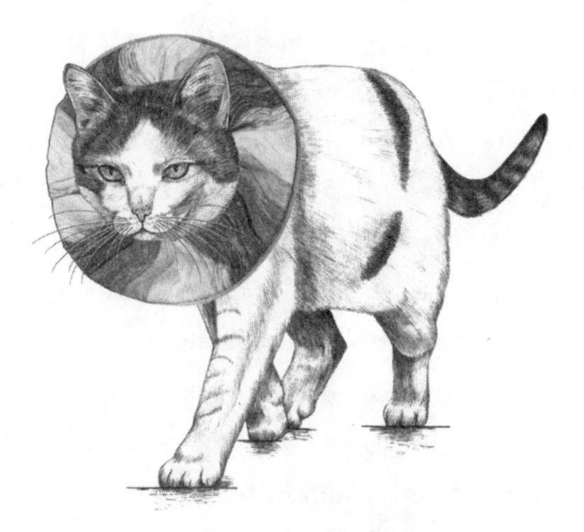

A cat wearing a colorful collar to alert potential prey.

Some reports suggest that the flouncy ruff is effective. Cats, however, are not big fans, no doubt because the neckwear is uncomfortable, but also perhaps because they realize they look ridiculous.

The alternative approach is to reduce the urge to hunt. Perhaps cats hunt just for the mental challenge, cat versus nature. If this is the case, maybe mental stimulation of other sorts can provide a sufficient cognitive workout.

Pet stores are full of games and puzzles. In one of my favorites, a treat is placed in a little well and covered; Nelson has to push aside a plastic leaf or move a round covering along a groove to expose the hidden kibble. In another, he has to paw a suspended cylinder so that it turns upside down and the food falls out into a terrain with many tall plastic spikes, making it difficult for him to reach with his paw to extract the goodie. The primary goal of these games is to keep cats from getting bored. But as an added benefit, could this cerebral stimulation be enough to sate the cat's desire to hunt?

Alternatively, the urge may be physical rather than mental—perhaps a cat has a need to jump, pounce, and bite, to move his muscles in particular ways. There are, of course, many appropriately designed toys for this purpose, and they get an enthusiastic response. Maybe enough physical exercise of this sort can substitute for hunting.

And finally, perhaps cats have a drive of a completely different type: they may go out on patrol because there is something missing in their diet, something they can only get from eating animal flesh. This idea is not unreasonable. Cats have very stringent needs based on their ancestral fare of eating almost nothing but meat. Although animal nutritionists have established standards to satisfy feline dietary requirements, not all commercial products meet them. If, in fact, a cat is hunting to obtain needed nutrients, then feeding the cat a well-designed cat food high in meat might lessen the desire to catch dinner on the wing.

The amount of money that some concerned cat owners will spend on these options is nearly unlimited. I know because I have. But what's the evidence that any of these interventions are successful in decreasing a cat's desire to hunt outside? For the most part, there is very little.

A team of British scientists recently decided to change that. They conducted a well-designed experiment to test the different ideas I've just mentioned, investigating whether any of these approaches reduced the number of prey brought home.

The researchers enlisted 355 cats from 219 households for the study. For seven weeks, before anything else happened, participants recorded every prey item their cats brought home. This was the baseline data to compare with what happened later. Then the cats were split into six groups; five experienced different interventions, plus a sixth control group for which nothing changed. For the next five weeks, the owners again tallied the prey brought home. The idea was to investigate whether there was a change in weekly predation rate after the interventions occurred.

Some of these ideas actually work! The most effective was switching cats to high-quality food prepared from fresh meat and without any grain. Cats receiving this diet substantially decreased the number of rodents and birds brought home.*

Playing with cats using toys that simulate hunting led to a thirty-five percent reduction in the number of mammals (rabbits and wood mice) brought home, but had little effect on bird predation. This result suggests that such play satisfied the cats' urge to hunt small mammals, but not the desire to go after birds. This explanation makes sense given that the procedure used in the study involved dragging a feather along the ground and, once the cat grabbed it, substituting a faux mouse he could kick, scratch, and bite to his heart's content. Perhaps this procedure was a good emulation of rodent, but not bird, hunting.

Nelson's favorite toy is a long pole with an object dangling on a string attached to the end. Nelson and our other cats will chase this endlessly, often jumping to snag the wiggling object in midair. I've often worried that perhaps I'm training the cats to catch birds, but this study suggests the opposite, that such play may satisfy their predatory urges. Hopefully, researchers will follow up with a study using toys of this type to see if they reduce bird hunting.

Cats wearing colorful collars produced the converse result: they brought home many fewer birds, but mammal returns dropped only slightly. This difference is easily explained. Small mammals do not

---

*More evidence that a vegan diet is not appropriate for cats! An alternative explanation is that the cats caught the same number of prey but ate or abandoned more of them outside rather than bringing them home. There's no reason to think this would occur, but the researchers acknowledge they can't rule out this possibility.

rely on vision, especially color vision, as much as birds, and so may have been less likely to notice the clownish cats heading their way.

Not all interventions reduced cat hunting, however. Wearing a bell had no effect, and playing with puzzles actually increased rodent predation by a whopping twenty-seven percent for reasons unknown (perhaps the games revved up the cats' mental machinery, whetting their appetite for more mental stimulation, or helped hone some key skill—your guess is as good as mine).

Overall, these results are good news (except for cat bell manufacturers): for people unwilling or unable to keep their cats inside, there are ways to reduce the toll on local wildlife.

THERE'S A LOT OF CONCERN ABOUT the impact of outdoor cats on the environment. Of course, pet cats, the topic of this chapter, are only part of the problem—we've already seen that predation rates by unowned cats are substantially higher than those of pet cats.

Outdoor cats affect wildlife—and humans too—in another way: they spread disease. Far and away the best known and most problematic is toxoplasmosis, a disease caused by *Toxoplasma gondii*. This one-celled organism can infect many different species, but it reproduces in only one type of animal: felines. When infected cats poop, they broadcast the parasites into the environment. Any animal unfortunate enough to ingest food or water containing *T. gondii* risks becoming infected; deaths have been reported in a number of endangered species, worrying conservationists. Humans, too, are at risk, and we are now learning that the consequences to us may be greater than previously realized. The extent to which toxoplasmosis transmitted by domestic cats is an environmental and human health problem is currently receiving considerable attention, but as of yet, it's not clear how big a problem this is. We also don't know the relative importance of owned and unowned cats in contributing to the problem, though, of course, indoor cats do all their pooping inside so they're not part of the problem.*

The debate about whether outdoor unowned cats are a problem and, if so, what to do about them is complicated and contentious, as much about sociology, ethics, and politics as it is about biology. Initially I thought I'd tackle these issues here, but I soon realized that one could write an entire book on the topic—indeed, several people have. Maybe someday I will as well. For this book, however, I could not do justice to the topic, and a few-chapters-long treatment would make no one happy.

For that reason, I'll limit my comments here to three points. First, with regard to the impact of hunting, unowned cats are responsible for the lion's share of the feline predation toll. The best estimate is that in the United States, unowned cats kill nearly two and a half times as many birds and nine times as many mammals as do pet cats. Research in Australia reveals a similar disparity.

Second, most of the vitriol has centered on what to do about unowned cats. Conservationists and cat advocates both care about animals, so you'd think they'd be able to work together, but in fact, they're usually at loggerheads. Everyone agrees that the fewer outside cats there are, the better, but the proposed solutions are radically different and compromise has been elusive.

Finally, third, at least for pet cats, there are uncontroversial options that can help, as we've seen in this chapter. For many reasons,

---

*People can get a variety of other diseases from cats as well, such as cat scratch disease and toxocariosis. Just as I was finishing this book, the first confirmed case of a cat transmitting COVID-19 to a person was published in the scientific literature. However, such transmission appears to be very rare.

keeping cats indoors is better for them and for the environment. For those who believe that cats should experience the outdoors, or those who—like me—don't have the willpower to stand up to a demanding feline, there are still things that can be done to minimize their impact.

THE NORTH AFRICAN WILDCAT is exquisitely adapted for inhabiting savannas and deserts, but not so much for life in kitchens and living rooms. I started this book by noting that the housecat is only semi-domesticated, that it hasn't diverged all that much from its wildcat ancestor. Most domesticated species, by contrast, have evolved to a markedly greater extent. Maybe it's time for us to help cats move further down the domestication road, to use the power of selection to create a cat well suited for modern living.

Twenty

The Future of Cats

What's next for *Felis catus*? The last few millennia have been quite a roller coaster for cats: worshipped as gods, massacred as satanic sidekicks, now arguably the world's most popular companion animal.* Along the way, the haughty North African wildcat has spawned a mélange of felines unlike anything the world has ever seen. Can the world of cats get any better?

I humbly submit that I've got a great idea.

*Whether there are more dogs or cats in the world depends on whom you ask. Worldwide, numbers are very fuzzy—there are probably more than half a billion of both.

. . .

WHEN I FIRST LEARNED of the existence of Savannahs, I was astonished. Not at the ability to cross a domestic cat and a serval (though that was surprising enough), but at the price people were willing to pay for one. And the more I thought about it, the more I became convinced that this was my chance—finally—to make my fortune at zoology. If people are willing to plunk down twenty grand for a long-legged cat with spots, just imagine what they might fork over for something truly extraordinary!

What would knock your socks off as the most amazing trait of another cat species to import into the domestic cat? Really, there's no question. Look no further than the Pixar movie *Ice Age* and its feline star. With daggers the size of bananas, Diego and his fellow saber-toothed tigers* were as cool as it gets, renowned from La Brea to Bedrock.† Who knows how much you could get for a saber-toothed housecat? Ka-ching!

Hear me out! Scimitar-fanged felines evolved in prehistoric times at least three times in cats and their relatives, as well as a fourth time in a large catlike marsupial in South America.‡ Such convergent evolution suggests that saber teeth—long, compressed, and sharp-edged, often with steak-knife serrations—can't be that difficult to evolve.

Now, I know what you're thinking. Who in their right mind would pay good money for a bloodthirsty killer? But you've got it wrong.

---

*"Saber-toothed cat" is more accurate. As we saw in chapter five, sabercats and tigers are distantly related.

†Referring, of course, to the hometown of Fred and Wilma Flintsone, who had a pet saber-toothed cat named Little Puss (not to be confused with their other pet, Dino, apparently a relative of *Brontosaurus*, which, of course, was ridiculous because dinosaurs were gone long before humans evolved).

‡*Thylacosmilus atrox.*

Smilodon fatalis, *a saber-toothed cat.*

Those prehistoric beasts were larger than lions, whereas we're talking lap cats. And anyway, no one knows what they did with their daggers—some paleontologists think they were scavengers because the teeth were too unwieldy to kill prey. In any case, this is beside the point. To be dangerous, the puss would have to be aggressive; the trick would be to create loving, placid, non-aggressive, saber-toothed tabbies.

So HOW ARE WE GOING to do it? Let's start with the new-school approach: genetic engineering. You may have heard of CRISPR, the new genomic tool that allows researchers to "edit" the genes of an organism, modifying the version of an existing gene or inserting a new one. Since its development a decade ago, CRISPR methods have been refined and advanced, and gene-edited individuals have been produced in many species, including—just to name some mammals—pigs, dogs, mice, cattle, opossums, and monkeys.

In theory, all we would have to do is use CRISPR to introduce the gene or genes responsible for saber teeth into a housecat. I won't go into the details, but doing so involves harvesting eggs from a female cat (just as is done in assisted fertilization clinics for people), fertilizing the egg with cat sperm, and at some point in the process, inserting the edited gene. Then the fertilized embryo is implanted into a queen.

The big challenge, of course, is to locate the genes responsible for producing saber teeth. Unfortunately, there are no living saber-toothed species to study. Some have suggested that the clouded leopard is dentally similar to sabercats. In reality, its canine teeth,* though longer than those of other living felines, are neither flattened nor anywhere near the length of sabercat blades. Studying the genetics of clouded leopards might identify genes involved in developing bigger teeth, but there's no guarantee that those genes are the ones we need to produce saber teeth.

But why even bother with a slightly long-in-the-tooth modern cat when the real thing lived not that long ago? *Smilodon fatalis* prowled the hills of Los Angeles until about ten thousand years ago. If Claudio Ottoni can get DNA out of similar-aged bones of domestic cats from Romania, perhaps scientists can do the same for fossil saber-toothed cats.

And indeed, they can. Scientists have now sequenced the entire genomes of two sabercat species, one from Chile, the other from the Yukon Territory.†

That's the easy part. As I remarked in chapter fifteen, genomes don't come with an index saying where to look for the genes respon-

*I can't help but remark on the unfairness of the sharp-and-pointies being named for dogs, rather than cats.

†Incidentally, comparison of the sabercat DNA to that of modern cats confirms the conclusion from the fossil record that the two branches of the feline evolutionary tree diverged about twenty million years ago.

sible for a particular trait. Now that we have the genome of saber-toothed cats, we have to figure out where among the more than two billion base pairs of DNA is the gene or genes responsible for canine elongation, flattening, and serration.

Studies on mice, humans, and other species have identified genes responsible for many aspects of tooth size and shape—checking those candidate genes out in sabercats would be a good place to start, but it won't be easy to figure out which ones are the key. Most likely the crucial bits of DNA are in regulatory regions that control the activity of other genes. One particularly promising gene is *Fgf10,* which is thought to play an important role in determining tooth sharpness. But there could be many others. Once the corresponding versions of the candidate genes are located in the sabertooth genome, researchers will need to introduce them into cat embryos separately and in combination to see what the resulting cats look like (the research probably would start by experimenting on mice, which are model organisms for such studies). It would be very hit-and-miss, and there's a decent chance that this approach wouldn't identify the right gene.

And I haven't even mentioned another key point: saber-toothed cats differ from regular felines in more than their trademark choppers. The skull, jaws, and musculature are modified in many other ways to accommodate the large teeth and the different means of biting. Probably there are many genes involved in the making of a sabercat, and all those genes would be important to identify.

Sadly (or not, depending on your view), genetic engineering is not going to produce a saber-toothed pussycat anytime soon.

THAT'S NOT TO SAY that genetic engineering won't play a role in feline futures. For already-identified genes, it may help produce a new and improved kitty. And one particular opportunity is obvious: cat allergies.

Worldwide, twenty percent of people are allergic to cats. For many, exposure to cats causes itchy eyes, sneezing, and congestion. For some, however, the reaction can be much worse, often leading to severe asthma attacks that require a hospital visit.

Such allergies are the result of a protein that occurs in cat saliva—infelicitously referred to as "Fel d 1" (pronounced "fell-dee-one").* When cats groom themselves, the protein gets on their skin and fur, then dries and flakes off in dander, joining the dust mites occupying our homes.

Cats vary greatly in how much of the protein they produce; cats that make less cause a weaker allergic response. Certain breeds—such as the Siberian and the Sphynx—are said to be hypoallergenic, but the evidence supporting these claims is weak. It may be that some individual cats don't produce much of the protein, but there's a lot of variability from one cat to the next, even within these breeds.

Allergic people will pay a lot of money to decrease their suffering without having to get rid of their cat. Fifteen years ago, a company capitalized on this opportunity by announcing that they had succeeded in breeding a line of low-allergen cats, which they sold for thousands of dollars each. The company subsequently failed for reasons that were never disclosed amid widespread accusations of fraud and a landslide of lawsuits from still-sneezing customers.

More recently, Nestlé Purina developed a cat food that is said to reduce the amount of allergen a cat produces by nearly fifty percent.

---

*Short for *Felis domesticus* Allergen 1. Actually, there are at least eight different human allergens that cats produce, but Fel d 1 is far and away the major problem.

The kibble contains a protein found in eggs that binds to Fel d 1 and prevents it from attaching to the molecules in the body that trigger an allergic response.

Of course, even better than a cat that produces low amounts of Fel d 1 would be a cat that didn't produce any at all. And that's where genetic engineering comes in. All scientists have to do is use CRISPR to replace the normal version of the gene that produces the Fel d 1 protein with a substitute allele that doesn't work.

Research is already well under way. The gene responsible for producing Fel d 1 was identified in 1991. Researchers at a company, Indoor Biotechnologies, subsequently made a non-functional version of the allele and have been successful at editing it into cat kidney cells in laboratory petri dishes. Now the question is whether they can get the procedure to work in actual cats. My bet is that they will eventually succeed (though it's notable that cats aren't yet among the many mammals for which gene-edited individuals have been produced).

Allergen-free CRISPR cats could be made in several ways. One approach would be to edit the gene in such a way that it is passed on to subsequent generations. This would be accomplished by injecting the edited gene into sperm or egg cells that would then be transmitted to the next generation. The result would be a new breed of cats.

It's unlikely that a company would take this project on as a service to humanity; rather, the corporate brass would have to think they could make money with the technology. To do that using this approach, they'd have to get into the business of breeding the cats. And once they sold an allergen-free cat, how would they make money on the allergen-free kittens that cat would beget? Even if they only sold neutered cats, others could sequence their genome, find the edited gene, and do their own CRISPR-ing. Perhaps the company would be able to retain intellectual property rights and demand a royalty on every allergen-free cat ever produced, akin to what Big Ag does with genetically modified seeds. Sounds like a bureaucratic and legal mess.

Not surprisingly, the company doing this research is taking an alternative route, hoping to develop a method of injecting the altered gene only into the glands that make the protein, thus eliminating Fel d 1 production. Because the altered gene wouldn't occur in the sperm or eggs, it wouldn't be passed on to the next generation. Such gene therapy would allow the company to continue selling the product to each consumer—treating each cat (or her vet) as a separate customer—without having to try to retain control of the offspring of cats whose genes they edit.

Of course, there's the question of animal welfare: will eliminating the gene have other repercussions for the cat? That is, does the protein serve some beneficial purpose for cats such that eliminating it would cause health problems? The fact that cats vary so greatly in how much protein they produce suggests that the protein can't be that important. If it were, then cats with high levels of the protein would be much healthier than low-protein cats, but there's no evidence of any such difference. On the other hand, Fel d 1 levels are generally highest in unneutered toms, so maybe the protein has something to do with reproduction in male cats. What that something is, and whether it really matters to the cat, remains to be seen.

BUT ENOUGH SNIFFLING about allergies; let's get back to the quest for the saber-toothed housecat. If genetic engineering is out, that leaves old-fashioned artificial selection. At first, that seems like an impossibility: how can we sculpt curved scimitars out of the dental battery of modern cats—they don't look at all like miniature sabercats! But recall that ancestral Persians and Siamese looked nothing like their current smooshed and angular selves. That's the power of selection.

The first thing we'd need to do is find some long-in-the-tooth cats, by which I mean cats with particularly long canines. As far as I'm aware, no one has ever studied variation in eyetooth length in domestic cats, but I'll readily admit that I'm not up on the feline orthodontic literature.

Naively, what I assumed we'd do is go to animal shelters and the like and measure the teeth of the cats therein. We'd then adopt the cats with the longest teeth—assuming they weren't neutered—and take them home to start our breeding colony.

I've had this idea for a long time now, but one stumbling block was the problem of what to do with all the surplus cats. For artificial selection to work, we'd need to produce a lot of kittens. From this bounty, we'd choose the cream of the crop, the crème de la crème of toothiness, and allow only those exceptionally endowed felines to breed. We'd do this generation after generation. But what about all the others, what would happen to them? My thought was that they wouldn't have a happy ending.

Fortunately, however, I've learned a way around this problem, thanks to Judy Sugden and Karen Sausman. Early in the process, we'll base our selection not just on the length of their teeth but also on their disposition. By choosing the cats with the best combination of fangs and friendliness, we'll end up with kittens that will be easy to give away if their sabers aren't up to snuff.

As we proceed, we'll have to be careful that the cats are healthy and their teeth functional. Because the skulls of sabercats differ from those of other felines to accommodate the sabers and the large muscles needed to use them, we may have to add other traits to our selection process to produce a cat whose parts fit together well. No doubt there will be unexpected hiccups. But my guess is that if breeders could produce a cat lacking a nose in a couple of decades of selection, then developing a saber-toothed housecat is likely to be successful. The fact that natural selection has done it at least four times in mammals suggests that it can't be that hard.

A saber-toothed housecat.

I've shared this idea with students in my classes on evolution and mammalogy over the last decade. A way to get rich quick, I told them—just remember me when the profits come rolling in. But no one has taken me up on the idea, so now I'm sharing it with you, my dear reader. Good luck, and please keep me in mind for one of your first saberkittens.

IN THE MEANTIME, as we await the development of the new Smilodon® breed, artificial selection will no doubt continue in other ways. Enthusiasts will continue crossing breeds with different traits to create new, ever-more-bizarre combinations (a Lykoi-coon, anyone?).

Breeders will pounce on whatever new mutations arise, no doubt leading to breeds with novel hair colors, textures, and patterns. Why not a cat with its tail curled back over its head, like some dog breeds (if such a trait isn't harmful)? Who knows what new trait will catch the fancy of the cat fancy?

One aspect of catness that has been surprisingly underexplored is the development of breeds with a broader range of body size. Look at dogs, ranging from six-pound Chihuahuas to two-hundred-and-fifty-pound English Mastiffs.

Frankly, I'm surprised that no one has bred a larger cat. Maine Coons are substantially larger than the ancestral wildcat, but nowhere near the feline upper limit (the Siberian tiger weighs up to 650 pounds; some saber-toothed cats approached a half ton). Moreover, Mainers have always been big. I'm not sure how much larger, if at all, they've gotten in recent decades. The only large breed to arise in recent years is the Savannah, whose size is the result of its ancestral serval heritage. Maybe cat breeders are showing some unexpected common sense and not selecting for cats big enough that they could be dangerous.

On the other hand, selection for small size has occurred in recent times. The smallest recognized cat breed is the petite Singapura, weighing in at six pounds for males and four for females. "Teacup" Persians are said on the internet to be about the same size. Given the existence of the Asian rusty-spotted cat at barely half that heft, I suspect that smaller breeds of cats could be developed.

LET'S MOVE BEYOND idle speculation about what type of cat someone might want to develop. There's one new breed that truly is urgently needed.

Cat behavior guru John Bradshaw has the solution to the outdoor pet cat problem: we should breed cats that don't want to go outside or don't have the urge to hunt. His argument is simple: cats vary in their desire to hunt, and some of this variation is likely based on genetic differences. All we have to do is selectively breed the cats happy to stay indoors, or if they go out, the cats that have no interest in stalking prey. The result would be a bird-friendly breed of cat.

I'm not aware of any data on breed differences in desire to go outside, but certainly some breeds are more active than others, which might correlate with such inclinations. The survey of veterinarians mentioned in chapter twelve reported a spectrum of activity levels, from highly active Abys and Bengals to sedentary Persians and Ragdolls. Probably not coincidentally, the same rankings were given to cats' propensity to hunt birds when outside, though, surprisingly, non-pedigreed cats ranked even higher than Bengals. The survey of four thousand Finnish cats gave congruent results: nearly two thirds of Abyssinian, Bengal, and randombred cat owners strongly agreed that their cats liked to chase small animals, compared to less than forty percent of those living with both Persians and Ragdolls.*

These interbreed behavioral differences have arisen without breeders intentionally selecting for them; the traits must be affected by genes that are under selection for other reasons. I have no doubt that if breeders intentionally chose cats that showed no interest in going outside or hunting, they would quickly develop a breed with those characteristics.

---

*Comparable differences were reported in response to a question asking whether their cats get "excited" (e.g., makes chirping or chattering noises, or lashes its tail) when looking at birds or other small animals at the window."

This would truly be a breed for the twenty-first century: adapted for city living, friendly to the environment, sedate and laid back.

Does this mean that the contribution that pet cats make to the wildlife carnage problem will be solved? The answer is no for two reasons. First, many people prefer cats that are active and energetic, like Bengals and Abyssinians. My guess is that "desire to go outside" and "propensity to hunt" are traits closely linked to activity level. In biological terms, these traits are probably affected by the same genes. When traits are genetically correlated, it is difficult to select for one of the traits without getting an evolutionary response in the other as well. As a result, it will be quite a challenge to produce a cat with the energy of a Bengal that doesn't want to go outside and hunt. But maybe I'm wrong. Five percent of respondents in the Finnish survey said their Bengals didn't like to chase small animals. By breeding those animals, perhaps a new variety of Bengal could be developed that is energetic and active, but not inclined to hunt.

A bigger problem is that most people do not own pedigreed cats. And where do these people get their cats? By and large, not from other homes; the vast majority of owned cats in the United States are neutered. Rather, most new cats in a household are previously unowned feral or colony cats.

Let's think about how such cats may be evolving. Consider which outdoor cats are most likely to be caught, adopted into a home, and neutered. Those that are more sociable, more likely to hang out around people.

And which won't get caught and will remain outside, passing their genes on to the next generation? The ones that are scaredy-cats, afraid of people, able to survive by their own wits. This is natural selection! And this selection is favoring unfriendliness to humans. We would expect unowned cat populations to evolve to be warier and less friendly;

they certainly wouldn't be evolving to lose their tendencies to go outside and hunt.

The only relevant study supported this hypothesis. Examining the offspring of pet queens, researchers found that kittens sired by an unowned outside-living male were less friendly to humans than those fathered by unneutered pet cats (friendliness being measured by how long the kitten would sit in her owner's lap). This finding suggests that a genetic difference exists between unowned and pet males, the unowned males possessing genes that make them less friendly. The study, however, was based on a very small sample of cats; as far as I'm aware, no follow-up study has ever been conducted. Whether, in fact, unowned populations are evolving to become less friendly is an important topic for future research.

THE LIMITED RESEARCH on whether natural selection is operating on unowned cat behavior is emblematic of a larger issue: there are almost no data on how unowned cats are evolving, even in Australia, where so much other cat research has been conducted. This seems like a missed opportunity given the scale of ongoing research and conservation activities.

We might predict that evolution is occurring in two ways. One is simply that feral cats could be reverting to the lifestyle of their ancestral wildcats, reversing the effects of domestication. It's not hard to imagine these cats filling the same ecological niche as the North African wildcat and basically evolving back from whence they came.

Alternatively, *Felis catus* may be striking out in new evolutionary directions. Feral cats occur in many places where there are few or no large predators. Look at how the coyote has taken advantage of that situation, occupying every imaginable habitat and evolving much larger in size in the absence of wolves that used to keep them in check. Maybe feral domestic cats are doing the same.

Consider the vast array of different habitat types occupied by cats in Australia: scorching red deserts, cold and snowy temperate mountains, rainforests, and grasslands. As an evolutionary biologist, my first thought is that these cats must be adapting to the different circumstances they are experiencing. Desert cats have probably evolved ways to cope with heat and scarce water, mountain-dwelling populations in the south to the cold and snow. Different prey species require different hunting adaptations; different predators—dingoes, Tasmanian devils, big lizards—require different ways to escape. Walking on sand poses different challenges from climbing on boulder fields.

A quick survey of the variety of small-cat species around the world illustrates how cats adapt to different environments: margays with reversible ankle joints to descend trees headfirst, sand cats with the soles of their feet covered in hair to walk across deserts, fishing cats with webbed toes for the life aquatic, to name just a few. Even though Australian cats have been evolving for hundreds, rather than millions, of years, they may have started to evolve in some of these directions.

For the most part, we don't know if this is happening. No one has examined whether physiological adaptations have arisen to live in different climates, nor are there data on body proportions or other anatomical traits. Behavioral differences for living in different environments and coping with different predators and prey haven't been studied either.

One thing is certain, though: feral cats—in Australia or elsewhere—do not have any of the extreme traits seen in some cat breeds. I've never seen a photo of a feral cat with a smooshed-in face like a Persian, nor one with the elongated muzzle of the modern Siamese. No short legs, no hairlessness, no curly ears. The conclusion is obvious: if domestic cats of those breeds were abandoned outside, they either wouldn't last long or natural selection would eliminate their extreme traits from the populations in subsequent generations.

Many Australians are convinced that their ferals are particularly large. And who could blame them with headlines like "Giant Wild Cats Weighing up to 25 lb Are Roaming Central Australia"? Such articles appear now and again, accompanied by photos and videos that appear to show extremely large cats, including one estimated to be five feet long! Nonetheless, the data do not support these claims; with one dubious exception, all the scientific research indicates that Australian ferals are unexceptional in size.

The debate over the size of Australian feral cats is not entirely academic. A team of scientists succeeded in getting Australia to ban the importation of Savannahs on the grounds that if such large cats were to become established in the wild, they could have devastating effects on prey species too large to be tackled by typical feral cats.* My view is that if there were a niche for larger cats available in Australia, then the local ferals would already have evolved to get larger; the lack of evidence for such size increases suggests that larger size is not favored. This reasoning implies that should a Savannah head into the bush Down Under, its large-size genes would be selected against and weeded out of the population. Nonetheless, the chance that large feral cats do exist in Australia, as well as the possibility that I am wrong about what would happen if the Savannah arrived, means that we shouldn't be too cavalier in this judgment; for those reasons, it may be prudent to keep Savannahs out.

---

*The authors also focused on the climbing and jumping abilities of Savannahs and the habitat use of their ancestral servals to argue that Savannahs might prey on species in places where typical Australian feral cats don't occur.

Worldwide, there have been many reports of feral cats that are both large and jet black. In Australia, there have been hundreds of sightings, often leading to widespread speculation of escaped black leopards. In Scotland, black cats given the name the "Kellas cat" are likely to be hybrids between housecats and Scottish wildcats, and the same explanation (involving the European variety of the wildcat) may explain observations of large black cats in the Caucasus. Large black cats have also been reported in New Zealand, Hawaii, and elsewhere.

Perhaps the most intriguing possible case of feral cat evolution involves another place where large, black feral cats have been sighted. In this case, though, the claims have been substantiated.

The place in question is the island of Madagascar. Cats were introduced there centuries ago, and as on islands around the world, they have had a major detrimental effect on the native animals. Surprisingly, they can even kill lemurs. Most feral cats in Madagascar look like typical mackerel tabbies, albeit big ones—one study found feral males to weigh twelve pounds compared to eight-pound non-feral males in nearby villages.

But there's an unusual twist: there are two types of feral cats. The more typical ones, just described, and a second, even larger type, with long graceful legs somewhat reminiscent of a serval and a pinched, narrow face like a Chartreux. Not only are these fitoaty, as the people of northeastern Madagascar call them, larger than the mackerel ferals (though no exact measurements are available), but they live in different habitats, the fitoaty found in deep forest, the mackerels at forest edge and around villages. And there's one more difference—the fitoaties are entirely black, just like so many other forest-dwelling felines.

When these strikingly handsome cats first came to the attention of scientists, some thought they might be an unknown species of feline,

A fitoaty.

but DNA tests confirmed their identity as *Felis catus*. Researchers are currently investigating the evolutionary significance of this elegant forest cat, as well as its impact on the native fauna.

So *Felis catus* has many possible futures, all of which may transpire simultaneously: genetically engineered novelties, new breeds that are more placid and indoorsy, unowned cats staying the same or even becoming less friendly, and feral cats evolving new ways to be a wild cat.

Could these different cats become different species? Remember that two populations are different species if they can't or won't interbreed and produce fertile offspring—being anatomically or physiologically different isn't enough in itself.* But as we've already discussed,

*Dogs are an instructive example. Despite their longer history of selective breeding and the immense variety in shapes, sizes, and behaviors, they all seem perfectly happy to mate with each other. However, the largest and smallest breeds may have trouble producing fully fertile offspring: size differences may make mating physically difficult (as with servals and housecats), and the mismatch in size

a gene can affect many different traits. The genetic changes involved in developing a cat breed that is very placid and lives indoors might also affect mating preferences such that these cats would refuse to mate with rough-and-tumble feral cats. Or perhaps an edited gene might have unexpected consequences, like rendering the sperm or eggs incompatible with those of cats that don't have the edited gene. It's certainly possible that some of these changes could lead two groups of cats to be reproductively incompatible and thus different species.

But let's look on a longer time horizon. Eventually, humanity will get its act together and stop pillaging the environment. Unless we completely annihilate all life, ecosystems will recover. Evolution will produce new species. Global species richness will rebound.

This newly reconstituted biodiversity will be descended from the species that have survived our onslaught. We can hope that lions, tigers, and grizzlies will still be around. But no doubt *Felis catus* will be. Already 600 million strong and found just about everywhere in the world, (no longer) domestic cats will be a prominent member of future ecosystems.

More than that, their ubiquity likely will make them ancestors of the new species that will repopulate the world. Just as the first cat to arrive in South America diversified in the last three million years to produce ocelots, kodkods, Geoffroy's cats, and six other species, today's Australian ferals may produce an adaptive radiation of Australian

might cause problems for a female carrying an embryo that is way too small or large for a dog of her size (these ideas have been suggested, but I have not been able to find data supporting them). So a modicum of infertility may occur in matings between large and small dog breeds, but even in this case, their gene pools are not isolated because both can breed with medium-size breeds, and thus over several generations, alleles from small breeds can make their way into large breeds and vice versa. As a result, *Canis familiaris* remains a single species.

species. Cats on oceanic islands may go their own way, each population evolving to become a different, distinctive species.

Proailurus, the first known cat species that lived thirty million years ago, gave rise to lions, cheetahs, sabercats, and more. Time will tell whether *Felis catus* spawns an equally rich evolutionary lineage. I wouldn't bet against it.

I'M GOING TO GIVE NELSON the last word. And that word is "mi-raaaa-ow," the phrase he uses when he wants to let me know that he very much wants to go outside.

Predictable in retrospect, our plan to mollify his al fresco urges by occasionally letting him into the backyard has backfired spectacularly: the more he goes outside, the more he wants out.

Fortunately, I've learned a few tricks to make the situation manageable. First, I dress him in his handsome tiger-striped vest. He looks dashing in his orange-and-black jacket on top of his luscious sable coat. More importantly, the ensemble makes him highly visible and proclaims that he is not an unowned cat.

I thought that one up on my own. But using my newfound cat-science smarts, I also make him wear a cat tracker on his collar.

So out he goes, usually swinging by the back door once an hour for a pat, a drink, a treat, or all three. Then he heads back out, repeatedly through the day, sometimes spending as much as six hours outside. Meanwhile, I sit at my desk, writing this book or doing something else, every few minutes glancing at the tracker app on my iPhone to see where he is. Occasionally, he leaves the yard, either wiggling through a hole that some animal—a groundhog or raccoon, maybe one of the cats—has dug under the fence, or jumping up and over at a spot where the fence isn't that high.

Nelson out and about.

I don't mind him wandering through neighbors' backyards or nearby subdivisions, though I do wish he'd avoid the houses with big dogs. He seems to have learned his lesson at one of those: ever since the time I had to retrieve him out of a tree he'd been chased up by a couple of English Pointers, he's mostly stayed away from that residence. He still wanders through the other yard, though, even though two large Labradors periodically go tearing around it when they're let out.

So I mostly let him be as he wanders. He goes no more than about two hundred yards from the house. Most directions are fine, but what I don't like is when he goes up to the busy road fifty yards to the east. When he heads that way, I go on high alert, and when he approaches the street, I scramble up there to intercept him and bring him home, much to the amusement of my neighbors.

It can be particularly annoying to have to do that several times in a single day. My suspicion is that sometimes he's too tired or lazy to walk home. Instead, he thinks to himself, "I'll just walk up to the road and call an Uber." And sure enough, moments later I appear to scoop him up and transport him home.

This arrangement worked well enough during COVID's work-at-home times, but as the prospect of going back to my office and traveling out of town approached, I began to worry. Nelson now becomes very aggravated when he doesn't get to go outside, and he takes that aggravation out on the other cats. It seemed like we had boxed ourselves into a corner—our system worked well as long as we were home all day to supervise Nelson's jaunts.

Then the cat science gods smiled on me. While reading about research on the impact of cats on the native fauna Down Under, I learned that the Aussies have perfected cat-proof fences—the trick is to bury the bottom of the fence underground and have the top flop over so that cats can neither climb up nor jump on top of it. These fences allow the scientists to experimentally test the effect of cats on ecosystems by creating enormous enclosures, some with cats, others without.

If the researchers could keep cats from escaping multi-square-mile enclosures, surely we could keep Nelson in our suburban-size backyard. A quick google revealed that several companies have followed the lead of the Aussie scientists and now market cat containment systems. They're not cheap, though, so we decided to go the DIY route, a work now in progress. Fingers crossed that Nelson won't be able to outwit us.

Even if Australian ingenuity has provided a reasonable compromise between Nelson's yearnings for open sky and our concerns for his well-being, there's the issue of his impact on other species. Some of my conservationist colleagues are appalled that I have failed, yet again, to raise an indoor-only feline. And indeed, there's nothing that

Nelson seems to enjoy more than stalking birds. He's keenly aware of their presence in the trees, as they fly by, and especially on the ground. He lies concealed in a bed of hostas—I wouldn't know he was there if not for the tracker—fifteen feet from where doves feed on seed spilled from the bird feeder above.

But here's the good news: Nelson, so exceptional in so many ways, is an inept hunter. He gets the idea of staying low to the ground and behind cover, but he doesn't understand the concept of slinking forward, of getting within pouncing distance. As a result, his rushes always come up short, the birds flying away well before he gets remotely close. I'm fairly confident he's never succeeded in catching anything.

You could blame Nelson's lack of training as a kitten. But Jane and Winston, hand-raised from two weeks old, are quite adept at catching rabbits. So lessons from momcat are obviously not essential to developing hunting prowess. Although some research has been conducted on how cats learn to be effective predators, more is needed.

And that call for research is a fitting way to finish this book. There's so much we know about cats, but so much more we don't. From topics as disparate as what's going on inside their heads, what impact they have on North American wildlife populations, and where exactly they were domesticated, there's a lot left for us to learn. It's an exciting time to be an ailurologist!

Acknowledgments

I am extraordinarily grateful to so many people who went above and beyond to educate me on anything and everything related to cats. For extensive conversation, repeatedly providing me with information or materials or otherwise being incredibly helpful, I thank Kyler Abernathy, Jason Ahistus, Mike Archer, Adam Boyko, Gordon Burghardt, Cris Bird, Ben Carswell, Martina Cecchetti, Francesco Cinque, Mike Cove, Marion Crain, Mikel Delgado, Justin Dellinger, Chris Dickman, Josh Donlan, Carlos Driscoll, Lucy Drury, Martin and Amanda Engster, David Fite, Harry Greene, Sarah Hartwell, Kathi Hoos, Anthony Hutcherson, Craig James, Jukka Jernvall, Roland Kays, Allene Keating, Heidy Kikillus, Bryan Kortis, Karen Kraus, Karen Lawrence and the Feline Historical Museum, Mike Letnic, Katie Lisnik, Leslie Lyons, Kerrie Anne Loyd, Fiona Marshall, Jenni McDonald, Alexandra Newton McNeal, Hugh McGregor, Emily McLeod, Jill Mellen, Jane Melville, Jo-Ann Miksa-Blackwell, Kim Miller, Vered Mirmovitch, Katherine Moseby, Tom Newsome, Nicholas Nicastro, Peter Osborne, Claudio Ottoni, Marissa Parrott, Troi Perkins, Nancy Peterson, Tracie Quackenbush, Andrew Rowan,

John Read, Grace Ruga, Jill Sackman, Bob Sallinger, Karen Sausman, Gary Schwartz, Grant Sizemore, Eric Stiles, Molly Stinner, Judy Sugden, Agnes Sun, Katherine Tuft, Wim Van Neer, Keoni Vaughn, Angela Weatherspoon, Linda Winters, Melissa Vetter and the Washington University libraries (especially the Inter-Library Loan office), and Jill Gordon and the Saint Louis Zoo library.

In addition, for answering questions or helping me in a wide variety of other ways, I thank Dani Alifano, Jane Allen, Carissa Altschul, Stefano Anile, Erika Bauer, Luisa Arnedo Beltran, Andrew Bengsen, Rhea Bennett, John Boothroyd, Pavel Borodin, Bjarne Braastad, John Bradshaw, Bonnie Breitbeil, Brittani Brown, Sarah Brown, Linda Bull, Renee Bumpus, Scott Campbell, Loretta Caravette, Laura Carpenter, Alexandra Carthey, Linda Castaneda, Hollie Colahan, Dan Dembiec, Melissa Drake, Deborah Duffy, Lee Dugatkin, Mark Eldridge, Sarah Ellis, Bobby Espinoza, Zach Farris, Jillian Fazio, Rosemary Fisher, Jess Flaherty, Kerry Fowler, Mike Gillam, Abigail Gough, Andrea Griffin, Idit Gunther, Liz Hansen-Brown, Ben Hart, Anne Helgren, Alison Hermance, Salima Ikram, Sandy Ingleby, Candilee Jackson, Pat Jacobberger, Cathy Johannas, Norman Johnson, Pam Johnson-Bennett, Holly Jones, Ken Kaemmerer, Gail Karr, Scott Kayser, Tom Keeline, Susan Keen, Michael Keiley, John Keilman, Scott Keogh, Marthe Kiley-Worthington, Gwendolyn LaPrairie, Sarah Legge, Chris Lepczyk, Carolyn Lesorogol, Dan Lieberman, Kit Lilly, Travis Longcore, Daans Loock, Shu-Jin Luo, Doug Menke, Ashleigh Lutz-Nelson, David Macdonald, Peter Marra, Henry Martineau, Becca McCloskey, Karen McComb, Robbie McDonald, Brennen McKenzie, John Moran, Cheryl Morris, Desmond Morris, Pavitra Muralidhar, Asia Murphy, Autumn Nelson, Darren Niejalke, Kristin Nowell, Stephen J. O'Brien, Ken Olsen, Helen Owens, Craig Packer, Barb Palmer, Jim Patton, Diane Paxson, Bret Payseur, David Pemberton, Ben Phillips, Melanie Piazza, Danijela Popovic, Niamh Quinn, Nathan Ranc, Susanne Renner, David Reznick, Harriet Ritvo, Deborah Roberts, David Roshier,

Dion Ross, Julius Bright Ross, Manuel Ruiz-García, Craig Saffoe, Marty Sawin, Doug Schar, Martin Schmidt, Susanne Schötz, Karin Schwartz, Trisha Seifried, Lara Semple, Brad Shaffer, Lynne Sherer, John Smithson, Philip Stephens, Ann Strople, Mel Sunquist, Amy Sutherland, Lourens Swanepoel, Teresa Sweeney, Allene Tartaglia, Patti Thomas, Chris Thornton, Dennis Turner, Yolanda van Heezik, Jean-Denis Vigne, Richard Wang, Georgia Ward-Fear, Wes Warren, Jim Wedner, Lars Werdelin, Mick Westbury, Annette Wilson, Robbie Wilson, Leigh-Ann Woolley, Mindy Zeder, and Iris Zinck.

Lucy Drury, Megan Kasten, Carolyn Losos, Elizabeth Losos, and David Schanzer read much or all of the penultimate version of this book—their comments were extensive and very helpful.

Many thanks to Max Brockman, my agent, for helping develop the idea for this book, and to my editor, Wendy Wolf, for shaping the book and giving me so many pointers that always turned out to be right, despite my immediate reaction. Thanks, too, to Paloma Ruiz, Jane Cavolina for catching many embarrassing grammar and wording blunders and all-around good copyediting advice, and Lynn Buckley, Meighan Cavanaugh, Cliff Corcoran, and Jennifer Tait. Thanks, too, to Lynn Marsden for cat wrangling and photo taking. Dave Tuss's illustrations are everything I had hoped they would be; I am grateful to him for putting up with my ever-changing requests and suggestions.

The genesis for this book was a course I taught in the fall of 2016, "The Science of Cats." Many thanks to Harvard University for approving and supporting the course, and to the twelve wonderful freshmen who took it.

Finally, I thank my parents, Joseph and Carolyn Losos, for raising me in a house with cats and nurturing my curiosity. My mom was my biggest fan as I wrote this book and gave me invaluable notes on the penultimate version. Most importantly, I thank my wife, Melissa Losos, for putting up with years of incessant jibber-jabber about all matters feline, and for her love and support.

Notes on Sources

A much more extensive list of references and additional commentary can be found at the author's website, www.jonathanlosos.com/books/the-cats-meow-extended-endnotes. Except where otherwise noted, all website addresses were correct as of June 15, 2022.

CHAPTER ONE: THE PARADOX OF THE MODERN CAT

Differences and similarities in behavior of different species were reported in M. C. Gartner et al., "Personality Structure in the Domestic Cat (*Felis silvestris catus*), Scottish Wildcat (*Felis silvestris grampia*), Clouded Leopard (*Neofelis nebulosa*), Snow Leopard (*Panthera uncia*), and African Lion (*Panthera leo*): A Comparative Study," *Journal of Comparative Psychology* 128 (2014): 414–26. A post on *Scientific American*'s website reviewed the scientific literature and said that scavenging on humans was reported much more for dogs than for cats, https://www.nationalgeographic.com/science/article/pets-dogs-cats-eat-dead-owners-forensics-science; see also M. L. Rossi et al.,

"Postmortem Injuries by Indoor Pets," *American Journal of Forensic Medicine and Pathology* 15 (1994): 105–9. One of the world's leading authorities on domestication, the Smithsonian archaeologist Mindy Zeder, has written several good overviews that are excellent starting places on the topic, e.g., M. A. Zeder, "The Domestication of Animals," *Journal of Anthropological Research* 68 (2012): 161–90. Statistics on the percentage of pedigreed and neutered pets are from the Humane Society, https://www.humanesociety.org/resources/pets-numbers, accessed January 4, 2019.

CHAPTER TWO: THE CAT'S MEOW

The original source is Sarah Brown's PhD dissertation, "The Social Behaviour of Neutered Domestic Cats" (University of Southampton, England, 1993). Nicholas Nicastro kindly told me the backstory of his PhD research in email correspondence in January 2021. The paper reporting the results of his study on the different types of meows is N. Nicastro and M. J. Owren, "Classification of Domestic Cat (*Felis catus*) Vocalizations by Naïve and Experienced Human Listeners," *Journal of Comparative Psychology* 117 (2003): 44–52. The follow-up study, which included people who lived with the subject cats, was reported in S. L. H. Ellis et al., "Human Classification of Context-Related Vocalizations Emitted by Familiar and Unfamiliar Domestic Cats: An Exploratory Study," *Anthrozoös* 28 (2015): 625–44. Schötz's research is nicely summarized in her 2017 book, *The Secret Language of Cats: How to Understand Your Cat for a Better, Happier Relationship* (New York: Hanover Square Press). Schötz combines chirp and chattering into one category, chirp-chatter, but it seems to me that they are distinct sound types. Cameron-Beaumont studied some very cool and little-known felines in British zoos—the Asian wildcat, Geoffroy's cat, caracal, and jungle cat—and presented data on their rates of meowing, in C. L. Cameron-Beaumont, "Visual and Tactile Communication in the Domestic Cat (*Felis silvestris catus*) and Undomesticated Small Felids" (PhD dissertation, University of Southampton, England, 1997). Renee Bumpus, who has worked with sixteen species, confirmed that meowing is common in small cats, and usually used in mother-kitten or courtship contexts (email correspondence, January 15, 2021). Want to hear a wildcat species meow? Check out this serval: https://www.youtube.com/watch?v=Le1GEAHnaGo, or this caracal: https://www

.youtube.com/watch?v=mZ_CDMyz374. The zookeeper survey was published by Cameron-Beaumont et al., "Evidence Suggesting Preadaptation to Domestication Throughout the Small Felidae," *Biological Journal of the Linnean Society* 75 (2002): 361–66. Nicastro's work on African wildcat vocalizations was published in 2004, in "Perceptual and Acoustic Evidence for Species-Level Differences in Meow Vocalizations by Domestic Cats (*Felis catus*) and African Wild Cats (*Felis silvestris lybica*)," *Journal of Comparative Psychology* 118: 287–96. A nice newspaper article about this paper, from which Nicastro's "Mee-O-O-O-O-O-W!" quote is taken, appeared in the *Cornell Chronicle*, https://news.cornell.edu/stories/2002/05/meow-isnt-language-enough-manage-humans. The study on different types of purrs was published by Karen McComb; see McComb et al., "The Cry Embedded Within the Purr," *Current Biology* 19 (2009): R507–8. McComb provided some of the details on the experimental procedures in an email on February 3, 2021. Recordings of the solicitation call can be heard at https://www.cell.com/current-biology/supplemental/S0960-9822(09)01168-3#supplementaryMaterial.

CHAPTER THREE: SURVIVAL OF THE FRIENDLIEST

Chapter 1 of John Bradshaw's wonderful *Cat Sense: How the New Feline Science Can Make You a Better Friend to Your Pet* (New York: Basic Books, 2013) recounts reports from people who have tried to raise wildcats cats in the past. Nobel Prize–winning animal behaviorist Konrad Lorenz provided firsthand stories from his own experience in *Man Meets Dog* (London: Methuen & Co., 1954, p.16): "Anyone who has had the opportunity of knowing an African wild cat more intimately will have no doubt that no great effort would be needed to make a creature of this species into a domestic animal. In a way it is a born domestic animal." By contrast, he says, European wildcats are "completely untameable." The list of tamed species comes from E. Faure and A. C. Kitchener, "An Archaeological and Historical Review of the Relationships Between Felids and People," *Anthrozoös* 22 (2009): 221–38. Chapter 4 in *Cat Sense* has an excellent discussion of cat socialization. There isn't much research on this topic, so there is uncertainty when exactly the critical age for handling begins and ends. It might start as early as three weeks, for example. The first surveys of cat behavioral variation were published by D. L. Duffy et al., "Development and Evaluation of the Fe-BARQ: A New Survey Instrument for

Measuring Behavior in Domestic Cats (*Felis s. catus*)," *Behavioural Processes* 141 (2017): 329–41 (Deborah Duffy kindly provided the raw data); and for Finnish cats, S. Mikkola et al., "Reliability and Validity of Seven Feline Behavior and Personality Traits," *Animals* 11 (2021): 1991. A particularly interesting article on fetching behavior in cats by Sarah Hartwell can be found at the very useful Messybeast.com website (http://messybeast.com/retriever.htm), which I consulted frequently during this book's preparation. The differences between dogs and wolves are discussed in Brian Hare and Vanessa Woods's *The Genius of Dogs: How Dogs Are Smarter Than You Think* (New York: Plume, 2013). The extent to which the ability to follow a person's pointed finger is innate versus the result of growing up around humans is a topic of considerable debate. Compare Clive D. L. Wynne's *Dog Is Love: Why and How Your Dog Loves You* (Boston: Houghton Mifflin Harcourt, 2019) with Hare and Woods's *Survival of the Friendliest: Understanding Our Origins and Rediscovering Our Common Humanity* (New York: Random House, 2020), for example. There has been some research on whether oxytocin levels in cats increase after interacting with people, but those studies have not been published in the scientific literature, only reported on in the press (e.g., https://www.hillspet.com/pet-care/behavior-appearance/why-humans-love-pets). The tail silhouette study is summarized in J. Bradshaw and C. Cameron-Beaumont (2000), "The Signalling Repertoire of the Domestic Cat and Its Undomesticated Relatives," in *The Domestic Cat: The Biology of Its Behaviour,* eds. Dennis C. Turner and Patrick Bateson (Cambridge, UK: Cambridge University Press, 2000, pp. 67–93). An example of lion tail-raising can be viewed at https://www.youtube.com/watch?v=jAPd90ePJ_U).

CHAPTER FOUR: STRENGTH IN NUMBERS

D. W. Macdonald et al.'s fascinating chapter "Felid Society," in *Biology and Conservation of Wild Felids,* eds. D. W. Macdonald and A. J. Loveridge (Oxford: Oxford University Press, 2010, pp. 125–60), provides a wealth of information on topics related to social structure of felines. Kristyn Vitale, in "The Social Lives of Free-Ranging Cats," *Animals* 12 (2022): 126, nicely reviews the literature on social behavior in unowned cat colonies. I report densities in cats per square mile, but the study sites were often substantially less than a square mile in size (just as you can measure how fast you are driving in miles per hour without actually driving for an hour).

Information on Nachlaot cats from V. Mirmovitch, "Spatial Organisation of Urban Feral Cats (*Felis catus*) in Jerusalem," *Wildlife Research* 22 (1995): 299–310, and email exchanges with Vered Mirmovitch on February 19, 2021. Larger density estimates have been reported than those I quote, but they either refer to cats in confined areas or don't include the entire ranging area of the cats, thus artificially inflating the density estimate. Unlike the situation for Nachlaot's cats, we know what has happened in recent years on Ainoshima. Four decades after the initial studies, the village's human population has declined by more than half, perhaps because the fisherpeople are growing old and not being replaced by a younger generation. Feline tourism, with attendant feeding of the cats by visitors and residents, is increasing according to a newspaper article in 2014. As for the cats, they still seem to be subsisting primarily on fish scraps, and supplementation by the tourists' handouts hasn't been enough to prevent the population from declining by more than fifty percent. A. Mosser and C. Packer, "Group Territoriality and the Benefits of Sociality in the African Lion, *Panthera leo*," *Animal Behaviour* 78 (2009): 359–70, provide a nice review of the advantages of large prides for territory ownership. A good discussion of infanticide is in Pusey and C. Packer's chapter "Infanticide in Lions: Consequences and Counterstrategies," in *Infanticide and Parental Care*, eds. Stefano Parmigiani and Frederick S. vom Saal (New York: Routledge, 1994). Some of the details, though, are hard to read—nature red in tooth and claw can be pretty unpleasant. The evidence for the statement that survival of kittens is better when communally cared for is a bit thin. D. W. Macdonald's magazine article, "The Pride of the Farmyard," *BBC Wildlife* 9 (November 1991): 782–90, mentions it alongside a fascinating discussion of farm cat colonies. Kerby and Macdonald's 1988 chapter "Cat Society and the Consequences of Colony Size," in *The Domestic Cat: The Biology of Its Behaviour*, eds. Dennis C. Turner and Patrick Bateson (Cambridge, UK: Cambridge University Press, 2014), shows that kitten rearing is much more successful in groups that have better territories within a colony. D. Pontier and E. Natoli, "Infanticide in Rural Male Cats (*Felis catus* L.) as a Reproductive Mating Tactic," *Aggressive Behavior* 25 (1999): 445–49, review the occurrence of infanticide in domestic cats. The best reference about multiple females deterring infanticide is in Macdonald's *BBC Wildlife* article. Warner Passanissi, mentioned in the article, confirmed that his observations on infanticide suggest that groups of females are better able to prevent infanticide than a single female (email, April 8, 2021). Deer licking photos from P. Bisceglio, "Why Is This Deer Licking This Fox?," *The Atlantic*, October 23, 2017 (https://www.theatlantic.com/science/archive/2017/10/why-is-this-deer-licking-this-fox/543621/). The camera trap approach works particularly well when each cat has a unique coat pattern. Individuals of some species are harder to tell apart, however. For those individuals, even more complicated math is used to estimate the number of individuals based on how far an individual is likely to roam, and thus how likely it is to be detected on multiple cameras. Information on wildcats living near each other comes from letters written in the field by Willoughby Prescott Lowe, a specimen collector for the British Museum of Natural History, reported by John Bradshaw in *Cat Sense: How the New Feline Science Can Make You a Better Friend to Your Pet* (New York: Basic Books, 2013). The report on multiple paternity in cats from L. Say et al. "High Variation in Multiple Paternity of Domestic Cats (*Felis catus* L.) in Relation to Environmental Conditions," *Proceedings of the Royal Society of London B* 266 (1999): 2071–74. "Cardinal rule quote" from Bradshaw, *Cat Sense*, p. 211. I thank cat behavior guru Pam Johnson-Bennett (email, March 20, 2021) for pointing out how cat owners sometimes make the situation worse. The statement that interlopers can succeed in joining a colony comes from O. Liberg et al.'s chapter "Density, Spatial Organisation and Reproductive Tactics in the Domestic Cat and Other Felids," in *The Domestic Cat: The Biology of Its Behaviour,* 2nd ed., eds. Dennis C. Turner and Patrick Bateson (Cambridge, UK: Cambridge University Press, 2000), which indicates that female dispersal from one group to another occurs, but only rarely. In contrast, dispersal in males is more common.

CHAPTER FIVE: CATS PAST AND PRESENT

The evolutionary history of felines is well summarized in L. Werdelin et al.'s chapter "Phylogeny and Evolution of Cats (Felidae)," in Macdonald et al., *Biology and Conservation of Wild Felids.* Pet name popularity from a registry of cats with insurance policies ("Naming Your Cat," Nationwide Pet Health Zone, https://phz8.petinsurance.com/pet-names/cat-names/male-cat-names-1). Simba appears universally popular, with two other surveys placing the moniker first among male names and third among names for all cats, male and female. First Vet ("A Rover by Any Other Name," https://firstvet.com/us/articles/a-rover-by-any-other-name) reported 115 years of pet names from the Hartsdale Pet Cemetery in New York. J. R. Castelló's, *Felids and Hyenas of the World* (Princeton, NJ: Princeton University Press, 2020) is an excellent source for information on all species of cats. The dichotomy of "big" versus "small" cats ignores the fact that body size is a continuum in feline species—the intermediate-size lynx, caracal, and African golden cat eat intermediate-size prey. For information on phylogenetic relationships of living species, see W. E. Johnson et al., "The Late Miocene Radiation of Modern Felidae: A Genetic Assessment," *Science* 311 (2006): 73–77, and G. Li et al. (2016), "Phylogenomic Evidence for Ancient Hybridization in the Genomes of Living Cats (Felidae)," *Genome Research* 26 (2016): 1–11. Darwin, in *The Variation of Animals and Plants Under Domestication* (vol. 2, 2nd ed., published in 1883, pp. 292–93), citing the famous French zoologist Isidore Geoffroy Saint-Hilaire, who in turn was repeating what Daubenton published in 1756, stated: "The intestines of the domestic cat are one-third longer than those of the wild cat of Europe. . . . The increased length appears to be due to the domestic cat being less strictly carnivorous in its diet than any wild feline species; for instance, I have seen a French kitten eating vegetables as readily as meat." For discussion of cat veganism, see "Are Vegan Diets Healthier for Dogs & Cats?," Skeptvet, April 29, 2022 (https://skeptvet.com/Blog/2022/04/are-vegan-diets-healthier-for-dogs-cats/). M. A. Zeder's "The Domestication of Animals," *Journal of Anthropological Research* 68 (2012): 161–90 is a good review of domestication and brain reduction. African wildcat strut nicely shown in https://www.youtube.com/watch?v=iDiL4YSNxwc&t=249s. The website of Scottish Wildcat Action (https://www.scottishwildcataction.org/about-us/#overview) is a good place to start for information on this gorgeous cat.

CHAPTER SIX: ORIGIN OF SPECIES

Carlos Driscoll's work was published in "The Near Eastern Origin of Cat Domestication," *Science* 317 (2007): 519–23. Driscoll provided many details about his research by phone and email in conversations from December 2020 to June 2022. John Bradshaw repeatedly makes the point on hybridization affecting African wildcats in *Cat Sense*. A counterargument might be that Geoffroy's cat and related South American felines are extremely friendly as well, and they do not hybridize with domestic cats. If the South Americans could evolve such friendliness, why not African wildcats?

CHAPTER SEVEN: DIGGING CATS

M. A. Zeder, "The Domestication of Animals," *Journal of Anthropological Research* 68 (2012): 161–90, provides a good review of pathways to domestication. D. Réale et al. provide a review of research on animal personality, "Evolutionary and Ecological Approaches to the Study of Personality," *Philosophical Transactions of the Royal Society B* 365 (2010): 3937–46. Oriol Lapiedra's "Predator-Driven Natural Selection on Risk-Taking Behavior in Anole Lizards" appeared in one of the premier journals in the field, *Science* 360 (2018): 1017–20. I had advised him that the project would never work and was a poor use of his time, but he stubbornly refused to take my advice. Show's what I know! The risk to cats from dogs in human settlements was pointed out to me by my Washington University colleague Fiona Marshall. Working cat adoption programs are discussed in Jen Christensen, "Are Cats the Ultimate Weapon in Public Health," CNN, July 15, 2016 (https://www.cnn.com/2016/07/15/health/cats-chicago-rat-patrol). I thank Fiona Marshall for also suggesting to me the possible reasons why cat remains may be so scarce in archaeological deposits. Jaromir Malek's *The Cat in Ancient Egypt* (London: British Museum Press, 1993) is the go-to source for information on Egyptian cats, full of photographs, illustrations, and interesting stories. D. Engels's broader *Classical Cats: The Rise and Fall of the Sacred Cat* (London: Routledge, 1999) is also useful. Chinese cat archaeological remains first reported by Y. Hu et al., "Earliest Evidence for Commensal Processes of Cat Domestication," *Proceedings of the National Academy of Sciences* 111 (2014): 116–20; a follow-up paper identifying the species is Vigne et al., "Earliest 'Domestic' Cats in China Identified as Leopard Cat (*Prionailurus bengalensis*)," *PLoS One* 11, no. 1 (2016): e0147295.

CHAPTER EIGHT: CURSE OF THE MUMMY

An entertaining summary of Ottoni's work, with some interpretations with which I disagree, can be found in Ben Johnson's "How Cats Took Over the

World," June 20, 2017 (https://natureecoevocommunity.nature.com/posts/17958-how-cats-took-over-the-world). My description of Ottoni's work was enlightened by Zoom and email conversations, mostly in December 2020 and January 2021. A great introduction to animal mummies is the 2005 book *Divine Creatures: Animal Mummies in Ancient Egypt* (Cairo: American University in Cairo Press). Its editor, Salima Ikram, told me that dogs were the most mummified animal in Egypt in an email on December 30, 2020. Murder conviction details can be found at https://en.wikipedia.org/wiki/Murder_of_Shirley_Duguay. M. Golab strongly advocates for the out-of-Turkey viewpoint and provides a nice perspective on how Ottoni's paper can be interpreted in "Egyptian Cats, Anatolian Cats and Vikings: Separating Evidence from Fiction About the Cat Domestication," June 25, 2017 (https://www.anadolukedisi.com/en/cat-domestication-fiction-evidence/). Ottoni's findings were actually more complicated: not only did European wildcats not occur in Turkey, but they did not even have sole possession of the European continent. As expected, all of the Western European samples from earlier than 800 BC carried European wildcat DNA. But old samples from southeastern Europe, predating the appearance of farming in that area, told a different story: a ten-thousand-year-old bone from Romania turned out to have North African wildcat DNA, as did the remains of an eight-thousand-year-old puss from Bulgaria. Overall, half of the cat remains from sites in southeastern Europe that dated to five thousand or more years ago contained North African wildcat DNA. These findings were intriguing because the standard view was that the geographic ranges of the two wildcat subspecies did not overlap; these data suggest, to the contrary, that both subspecies were present in the region. When Ottoni compared the DNA of the North African wildcats from Europe with that of equally ancient cats from Turkey and nearby areas, he detected a slight but consistent genetic difference: a particular allele was common in Turkey but did not occur in Europe. This difference suggests that the North African wildcat populations in the two areas had been evolving separately for an extended period of time, long enough for genetic differences to accumulate. In other words, North African wildcats had been in southeastern Europe for quite some time, far in advance of the migration of humans from Turkey that brought agriculture to Europe. The subsequent appearance in Europe of the Turkish allele indicated an influx of Turkish cats to the European continent. Fiona Marshall provides an excellent discussion and synthesis of the implication of Turkish alleles showing up in southeastern Europe in "Cats as Predators and Early Domesticates in Ancient Human Landscapes," *Proceedings of the National Academy of Sciences* 117 (2020): 18154–56. Many sources—online, in books, even in scientific papers—say that the Egyptians called the Phoenicians "cat thieves." However, I have been unable to find an authentic primary source documenting this statement. My guess is that someone, at some time in the past, invented the epithet and now it's become accepted as fact. Similarly, statements about Egyptian efforts to repatriate cats taken out of Egypt all trace back to the writing of the Greek historian Diodorus Siculus, who wrote in the *Library of History,* Volume 1 (https://penelope.uchicago.edu/Thayer/e/roman/texts/diodorus_siculus/1d*.html): "And if they happen to be making a military expedition in another country, they ransom the captive cats and hawks and bring them back to Egypt, and this they do sometimes even when their supply of money for the journey is running short." Many published reports mischaracterize what he wrote, going so far as to sometimes say that armies were expressly sent out to invade other lands to capture the cats. Again, the explanation is probably that one writer got it wrong, and others have just copied the mistake, sometimes embellishing, and thus inaccuracies become repeated and eventually accepted as fact. The discussion of the geographical spread of cats was drawn primarily from D. Engels, *Classical Cats: The Rise and Fall of the Sacred Cat* (London: Routledge, 1999), and E. Faure and A. C. Kitchener, "An Archaeological and Historical Review of the Relationships Between Felids and People," *Anthrozoös* 22 (2009): 221–38. M. Toplak explained the revisionist history in Norwegian cat mythology in "The Warrior and the Cat: A Re-evaluation of the Roles of Domestic Cats in Viking Age Scandinavia," *Current Swedish Archaeology* 27 (2019): 213–45. More information on Ottoni's grant can be found in "A History of Cat-Human Relationship," University of Rome Tor Vergata (https://farmacia.uniroma2.it/a-history-of-cat-human-relationship-2mln-granted-by-erc-to-the-felix-project/).

CHAPTER NINE: CALICO TIGERS AND PIEBALD PUMAS

Information on the color of cats in art is primarily from D. Engels, *Classical Cats: The Rise and Fall of the Sacred Cat* (London: Routledge, 1999).

M. R. Clutterbuck's *Siamese Cats: Legends and Reality,* 2nd ed. (Bangkok: White Lotus, 2004) is an authoritative discussion of the *Tamra Maew* and modern-day cats from Thailand. Tortitude information from "'Tortitude' Is Real, and Other Fun Facts About Tortoiseshell Cats," October 2, 2018 (https://www.meowingtons.com/blogs/lolcats/tortitude-is-real-and-other-fun-facts-about-tortoiseshell-cats). A go-to reference, quoted by this website, is I. King, *Tortitude: The BIG Book of Cats with a BIG Attitude* (Miami: Mango Media, 2016). Why tortoiseshell and calico cats are almost always females is explained in Kat McGowan, "Splotchy Cats Show Why It's Better to Be Female" (https://nautil.us/splotchy-cats-show-why-its-better-to-be-female-3608/).

CHAPTER TEN: A SHAGGY CAT STORY

Toby's full name is Grand Champion of Distinction LapCats Tobias MacNifico. Cat popularity figures from the Cat Fanciers' Association, https://cfa.org/cfa-news-releases/top-breeds-2021/. Descriptions of Maine Coons appear in Frances Simpson's *The Book of the Cat* (London: Cassell and Company, 1903) and Harrison Weir's *Our Cats and All about Them* (Boston: Houghton Mifflin, 1889); photos of nineteenth-century Maine Coons are available at https://www.pawpeds.com/cms/index.php/en/breed-specific/breed-articles/forebears-of-our-present-day-maine-coons. One of the Norwegians who led the effort to develop the breed remarked in 1977, "Historically, little is known about the Norwegian Forest Cat. Most of what has been written about the cat and its background is supposition and pure guesswork. One thing is known however: the Forest Cat has existed in Norway for as long as can be remembered," (E. Nylund, "The Norwegian Forest Cat," *All Cats* 14: 27); http://www.skogkattringen.no/artikler/Norsk_Skogkatt_histore_m_bilder_2.pdf describes the initiation of the NFC breed, translated from Norwegian: "Collection of breed-typical individuals from all over Norway was made. . . . Cats were literally lifted out of the shed and set on the judge's table in hopes of fitting into the technical requirements which had been set. Some came through the eye of the needle, others did not. The breeding work could finally begin." According to Bjarne O. Braastad (email, June 2, 2021), the standard has not changed since these founding individuals were selected, meaning that the Norwegian Forest Cat breed today includes cats similar to those plucked out of the woods and off farms decades ago. More information is available at https://www.norgeskaukatt.co.uk /Norgeskaukatt/History/firstshowcats.html. The description of the Russian longhair is from Weir's book. I thank Lucy Drury (several emails beginning June 10, 2021) for walking me through the history of the Siberian and Drury (June 21, 2022) and Sarah Hartwell (August 19, 2022) for discussion of variation in the conformation of the same breed in different cat organizations. Irinia Sadovnikova's article on Siberians is very informative (https://www.pawpeds.com/cms/index.php/en/breed-specific/breed -articles/the-siberian-cat); in it, she described the origination of Siberians: "At that time the idea to create a Russian breed was up in the air. And of course it should have been called 'Siberian Cat,' due to the long history of this word collocation. But the appearance of this cat had not been obvious yet. It must have been semi-longhair, but what else? Type, size, shape of the head, muzzle contours, placement of ears—a wide variety of these features was represented in the urban and the suburban population of semi-longhair cats (let us call them 'conventionally aboriginal') studied by felinologists. They had to make choice on the basis of the predominant type in the population, taking into account already recognized breeds of semi-longhair cats, mainly the Maine Coon and the Norwegian Forest cat. Everybody wanted to refrain from the repetition of the existing things." Other useful articles are by A. V. Kolesnikov (http://www.tscharodeika.de/unmasked.html) and Olga O. Zaytseva (https://www.biorxiv.org/content/10.1101/165555v1); the latter expresses skepticism about the link between Siberians and Russian longhairs of the past. Pictures of street cats of Thailand are available at http:// www.siamesekittens.info/thailand.html. For Cairo, see Lorraine Chittock's *Cats of Cairo* (New York: Abbeville Press, 2000). For British cats, surf the internet or check out the BBC documentary *The Secret Life of the Cat*. Rather than relying on such studies, researchers should conduct a formal study of geographic variation in street cats. The feral cats on the nearby island of Lamu have a similar appearance to Sokoke Forest cats as well. Academy Award–nominated cinematographer Jack Couffer wrote a charming book, *The Cats of Lamu* (London: Aurum Press, 1998), about the cats occupying the old town on this picturesque Kenyan island. M. R. Clutterbuck's *Siamese Cats: Legends and Reality*, 2nd ed. (Bangkok: White Lotus, 2004, pp. 95–96) describes the history of the Burmese breeds.

CHAPTER ELEVEN: NOT YOUR FATHER'S CAT

A great resource on the characteristics of the different breeds are the breed pages on the Cat Fanciers' Association (https://cfa.org) and the International Cat Association (https://tica.org/) websites, such as the Persian breed page from CFA (https://cfa.org/persian/persian-breed-standard/). The controversy over acceptance of the Munchkin breed and its history can be found in "Munchkin: Fur Is Flying Over This Rare Cat Breed," *Tampa Bay Times,* June 14, 1995 at (https://www.tampabay.com/archive/1995/06/14/munchkin-fur-is-flying-over-this-rare-cat-breed/). An updated analysis of health problems, or lack thereof, can be found on an enthusiast's website, Munchkin Cat Guide, "Do Munchkin Cats Have Health Problems?" December 27, 2018 (https://www.munchkincatguide.com/do-munchkin-cats-have-health-problems/). The evolutionary history of felines is nicely summarized in L. Werdelin et al.'s chapter "Phylogeny and Evolution of Cats (Felidae)," in Macdonald et al., *Biology and Conservation of Wild Felids,* eds. D. W. Macdonald and A. J. Loveridge (Oxford: Oxford University Press, 2010).

CHAPTER TWELVE: INCESSANT JIBBER-JABBER

It's Show Time! by P. Maggitti and J. Anne Helgren (Hauppauge, NY: Barron's Educational Series, 1998) is a nice introduction to cat shows and *Meow: What Cats Teach Judges about Judging* gives the judge's perspective (K. J. Fowler, 2021), also pointing out that cat shows run by different organizations and in different countries vary from the format I've described. The survey of veterinarians was reported in B. L. Hart and L. A. Hart, *Your Ideal Cat: Insights into Breed and Gender Differences in Cat Behavior* (West Lafayette, IN: Purdue University Press, 2013). J. Anne Helgren's *Encyclopedia of Cat Breeds,* 2nd ed. (Hauppauge, NY: Barron's Educational Series, 2013) is my bible of cat breeds, with entertaining and thorough discussions of almost every cat breed . . . at least those in existence in 2013! Helgren's rankings of behavioral traits were the result of years of research as a writer for cat magazines and websites, involving attending cat shows; talking with breeders, owners, veterinarians; and "of course I interviewed the cats themselves" (email, November 16, 2020). Research on behavioral variation among breeds includes D. L. Duffy et al., "Development and Evaluation of the Fe-BARQ: A New Survey Instrument for Measuring Behavior in Domestic Cats (*Felis s. catus*)," *Behavioural Processes* 141 (2017): 329–41; J. Wilhelmy et al., "Behavioral Associations with Breed, Coat Type, and Eye Color in Single-Breed Cats," *Journal of Veterinary Behavior* 13 (2016): 80–87; and M. Salonen et al., "Breed Differences of Heritable Behaviour Traits in Cats," *Scientific Reports* 9 (2019): 7949. The Finnish study was by S. Mikkola et al., "Reliability and Validity of Seven Feline Behavior and Personality Traits," *Animals* 11 (2021): 9911. The idea that Persians were bred to be placid was suggested by Desmond Morris, *Catworld: A Feline Encyclopedia* (New York: Viking, 1996) and by Sarah Hartwell in an email (June 2, 2022); Rosemary Fisher proposed that such equability was necessary so that the long-haired cats could be groomed (email, July 27, 2022). The comparison of the Siamese breed cats to other cats in Thailand was made in M. R. Clutterbuck's *Siamese Cats: Legends and Reality*, 2nd ed. (Bangkok: White Lotus, 2004).

CHAPTER THIRTEEN: BREEDS OLD AND NEW

Harriet Ritvo's *The Animal Estate: The English and Other Creatures in the Victorian Age* (Cambridge, MA: Harvard University Press, 1987) provides an enlightening discussion of the development of the dog and cat fancies. Harrison Weir, the man credited with starting the fancy, described the variety of cats known at the time in *Our Cats and All About Them* (Boston: Houghton Mifflin, 1889). A very interesting parallel discussion of the derivation and antiquity of dog breeds can be found in a *New York Times* article by James Gorman, "How Old Is the Maltese, Really?," October 4, 2021 (https://www.nytimes.com/2021/10/04/science/dogs-DNA-breeds-maltese.html). I am indebted to Cris Bird for discussions of Siamese cat history (emails, June 19–21, 2021) that elaborated on the history she presents in "Sarsenstone Cattery: A Word About Siamese History and Body Type" (http://www.siamesekittens.info/siamhx.html) and "Sarsenstone Cattery: The Types of Siamese" (https://web.archive.org/web/20060930012742/http://home.earthlink.net/~sarsenstone/threetypes.html). A particularly interesting, if provocative, book on pedigreed dog breeding is Michael Brandow's *A Matter of Breeding: A Biting History of Pedigree Dogs and How the Quest for Status Has Harmed*

Man's Best Friend (Boston: Beacon Press, 2015). Conversation with Grace Ruga (October 8, 2021) provided the history of the American Curl; in a subsequent email (October 13, 2021), she emphasized that developing the American Curl was a team effort involving a number of others besides the Rugas. Also, although most of Shulamith's kittens were given away to friends and family, one was placed via a classified ad. Adam Boyko provided information on dog breeds (email, October 4, 2021). Details on the Virginia cats involved in the development of the Lykoi were provided by Patti Thomas (email, July 15, 2021). There is some controversy on how the Lykoi originated. The CFA and TICA breed pages, for example, provide different histories. See also Sarah Hartwell, "The Uncensored Origins of the Lykoi" (http://messybeast.com/lykoi-story.htm) for some unseemly details. A nice history of short-legged cats can be found in Hartwell's "Short-Legged Cats" (http://messybeast.com/shortlegs.htm).

CHAPTER FOURTEEN: SPOTTED HOUSECATS AND THE CALL OF THE WILD

Nether parts problems told to me by both Martin Engster (October 20, 2017) and Karen Sausman (July 21, 2021). Information about Savannah sizes and price can be found at https://www.savannahcatassociation.org/. To see a Savannah jump really high, go to https://www.youtube.com/watch?v=vprEInOl1oO. The destructive tendencies of bored Savannahs are discussed in T. David's *Savannah Cats and Kittens: Complete Owner's Guide to Savannah Cat and Kitten Care* (Canada: Windrunner Press, 2013); the merits of different generation Savannahs can be assessed at "Which Cat Is Right for You?" (https://savannahcatbreed.com/which-cat-is-right-for-you/). *The New Yorker* article, "Living-Room Leopards," by Ariel Levy, can be found at https://www.newyorker.com/magazine/2013/05/06/living-room-leopards. I conversed with Anthony Hutcherson on August 11, 2021 and November 19, 2020, as well as by email a number of times. A comprehensive list of attempted crossbreeds, "Domestic x Wildcat Hybrids" by Sarah Hartwell, resides at http://messybeast.com/small-hybrids/hybrids.htm. Karen Sausman's history and the Serengeti breeding program were detailed in phone and email conversations in November 2020 and later in 2021, as well as in a Zoo &

Aquarium Video Archive interview, available at http://www.zoovideoarchive.org/karen-sausman, the transcript of which was kindly provided by Loretta Caravette. I learned of the history of the Toyger by reading the cover article in the February 23, 2007, issue of *Life*, K. Miller, "Hello, Kitty: Inside the Making of America's Next Great Cat," and Alexandra Marvar's "You Thought Your Cat Was Fancy?," *New York Times*, May 27, 2020 (https://www.nytimes.com/2020/05/27/style/toyger-fever.html), as well as from conversations with Judy Sugden on November 22, 2020, and June 13, 2022, and in emails in between. The Darwin quote on Lord Rivers is from *The Variation of Animals and Plants Under Domestication* (vol. 2, 2nd ed., published in 1883, p. 221). The story of breeding twisty cats appears in R. Tomsho and C. Tajada, "Mutant Cats Breed Uproar; 'May God Have Mercy,'" *Wall Street Journal*, November 27, 1998 (https://www.wsj.com/articles/SB912119899369123000); discussion of twisty cats and the ethics of breeding for detrimental traits can be found in Hartwell's "Twisty Cats: The Ethics of Breeding for Deformity" (http://messybeast.com/twisty.htm). The MRI study of Persians' skulls is M. J. Schmidt et al., "The Relationship Between Brachycephalic Head Features in Modern Persian Cats and Dysmorphologies of the Skull and Internal Hydrocephalus," *Journal of Veterinary Internal Medicine* 31 (2017): 1487–1501. The UK study of health of three hundred thousand cats was published by D. G. O'Neill et al., "Persian Cats Under First Opinion Veterinary Care in the UK: Demography, Mortality and Disorders," *Scientific Reports* 9 (2019): 12952. A quick google will lead you to many websites explaining why declawing is a terrible idea. For example, see the Humane Society's page at https://www.humanesociety.org/resources/declawing-cats-far-worse-manicure, or the ASPCA's at https://www.aspca.org/about-us/aspca-policy-and-position-statements/position-statement-declawing-cats, or S. Robins's article "The Battle to Stop Declawing," May 10, 2021 (https://www.catster.com/lifestyle/the-battle-to-stop-declawing).

CHAPTER FIFTEEN: CATANCESTRY.COM

I spoke with Leslie Lyons by phone and email throughout spring and summer 2021. A good review of how scientists find the genes responsible for particular traits is in B. Gandolfi and H. Alhaddad, "Investigation of Inherited Diseases in Cats: Genetic and Genomic Strategies over Three Decades,"

Journal of Feline Medicine and Surgery 17 (2015): 405–15. Lyons nicely explains genetic testing for diseases in "Feline Genetic Disorders and Genetic Testing," Tufts' Canine and Feline Breeding and Genetics Conference, 2005 (https://www.vin.com/apputil/content/defaultadv1.aspx?id=3853845&pid=11203). Her "Everything You Need to Know About Genetics . . . You Can Learn from Your Cat!" ongoing series in *Felis Historica* provides many useful examples. The first six are available at the History Project, CFA Foundation, http://www.cat-o-pedia.org/articles.html. Another nice review of cat genetics by Pavel Borodin is available at https://scfh.ru/en/papers/cats-and-genes-40-years-later/. Lyons presented a very nice lecture on the P4 approach to feline health at the end of 2020: "Precision Medicine & Genomic Resources for Domestic Cats," Cornell University Video on Demand (https://vod.video.cornell.edu/media/Baker+Institute+virtual+seminar+series+-+Dr.+Leslie+Lyons/1_ouj3m230). An informative interview can be found at Growing Life, Our Feline Futures, episode 13, "Leslie Lyons: Exploring the Feline Genome" (https://www.gowinglife.com/leslie-lyons-exploring-the-feline-genome-our-longevity-futures-ep-13/). More information on the 99 Lives project is available at http://felinegenetics.missouri.edu/. An entertaining and enlightening discussion of the problem of lack of genetic variation in cat breeds can be found in L. Drury, "The Challenge of Diversity in Small Breeding Populations," *Cat Talk* 12, no. 1 (2022): 7–9. Carlos Driscoll suggested to me that colors seen in Cairo cats were probably not present in ancient Egyptian times in an email, May 21, 2021. Genetic differences among cats of the world are reported in a paper by S. M. Nilson et al., "Genetics of Randomly Bred Cats Support the Cradle of Cat Domestication Being in the Near East," *Heredity* 129 (2022): 346–55. A description of geographic differentiation is also found in the very different analysis of Alhaddad et al., "Patterns of Allele Frequency Differences Among Domestic Cat Breeds Assessed by a 63K SNP Array," *PloS One* 16 (2021): e0247092. Karen Lawrence's *The Descendants of Bastet: The Early History and Development of the Abyssinian Cat* (Canada: CFA Foundation and Harrison Weir Collection, 2021) should be consulted for a detailed history of the Abyssinian. H. G. Parker et al., "Genomic Analyses Reveal the Influence of Geographic Origin, Migration, and Hybridization on Modern Dog Breed Development," *Cell Reports* 19 (2017): 697–708, is a good source for background on dog breed evolution. A nice explanation of what cat genetic test results mean can be found in

Emilie Bess, updated by Karen Anderson, "We Tried the Top Two Cat DNA Tests and Here's What We Discovered," The Dog People, Rover.com (https://www.rover.com/blog/cat-dna-test/).

CHAPTER SIXTEEN: PUSSY CAT, PUSSY CAT, WHERE HAVE YOU BEEN?

According to the narration of the documentary, the density of cats is highest in the southeast of England. No reason was given for why Shamley Green in particular was chosen. At the time the documentary was filmed, most discussion of the behavior and ecology of *Felis catus* came from studies of feral cats or from cats living in colonies and varying in degree of human intervention from farm cats to those in managed colonies of strays; see, for example, chapters in *The Domestic Cat: The Biology of Its Behaviour*, 2nd ed., eds. Dennis C. Turner and Patrick Bateson (Cambridge, UK: Cambridge University Press, 2000); or in John Bradshaw, *Cat Sense: How the New Feline Science Can Make You a Better Friend to Your Pet* (New York: Basic Books, 2013). The Albany study was R. W. Kays and A. A. DeWan, "Ecological Impact of Inside/Outside House Cats Around a Suburban Nature Preserve," *Animal Conservation* 7 (2004): 273–83. Details on specific cats can be found at the BBC News website, "Secret Life of the Cat: What Do Our Feline Companions Get Up To?," June 12, 2013 (https://www.bbc.com/news/science-environment-22567526). In 2021–2022, even more cat tracker options were available and reviews were somewhat more consistent in their favorites. Here are four, but there are more: https://allaboutcats.com/best-cat-tracker; https://www.mypetneedsthat.com/best-cat-gps-tracker/; https://www.t3.com/us/features/best-cat-gps-tracker; and https://www.buskerscat.com/best-cat-tracker-uk. Information on the Cat Tracker project from an interview and subsequent emails with Troi Perkins in 2019 and 2020, as well as from Mark Turner, "Cat Tracker Project: How Cats Live Their (Nine) Lives," TechnicianOnline, January 25, 2016 (http://www.technicianonline.com/arts_entertainment/article_cc5adb66-c3e6-11e5-8a3a-3bc55bda07c1.html). The project website is also very informative: http://cattracker.org/tracks/, as is R. W. Kays et al., "The Small Home Ranges and Large Local Ecological Impacts of Pet Cats," *Animal Conservation* 23 (2020): 516–23. Australia and New Zealand data from H. Kikillus et al., "Cat Tracker New Zealand: Understanding Pet Cats Through Citizen Science," 2017 (http://cattracker.nz/wp-content/uploads/2017/12/Cat-Tracker-New-Zealand_report_Dec2017.pdf) and P. E. J. Roetman et al., "Cat Tracker South Australia: Understanding Pet Cats Through Citizen Science," 2017 (https://doi.org/10.4226/78/5892ce70b245a). The British study that reported cat mortality from cars is J. L. Wilson et al. (2017), "Risk Factors for Road Traffic Accidents in Cats up to Age 12 Months That Were Registered Between 2010 and 2013 with the UK Pet Cat Cohort ('Bristol Cats')," *Veterinary Record* 180 (2017): 195.

CHAPTER SEVENTEEN: LIGHTS, KITTYCAMS, INACTION!

See John Bradshaw in *Cat Sense: How the New Feline Science Can Make You a Better Friend to Your Pet* (New York: Basic Books, 2013 pp. 102–6) for more information on cat vision. Hernandez explained the background of the kittycam project in Mia Falcon, "UGA Researcher Studies Feline Feeding Behavior on Jekyll Island," *TheRed&Black*, September 20, 2015 (https://www.redandblack.com/uganews/uga-researcher-studies-feline-feeding-behavior-on-jekyll-island/article_8c5712e8-6714-11e5-93b4-9f3bcf7b738f.html): "Hernandez said the project was inspired by the multiple different species killed by her own domestic cat. 'When I started to pay attention about the scope of the impact of domestic cats on the environment, I was dismayed,' Hernandez said." National Geographic information from interviews with Kyler Abernathy, November 8, 2019, and November 11, 2019, supplemented by online articles such as https://www.nationalgeographic.org/article/creature-feature/. The kittycam videos can be seen at Hernandez Lab, University of Georgia, https://kaltura.uga.edu/category/Hernandez+Lab/33080331. I interviewed Kerrie Ann Loyd on August 30, 2019. Information on risky behaviors in K. A. Loyd et al., "Risk Behaviours Exhibited by Free-Roaming Cats in a Suburban US Town," *Veterinary Record* 173 (2013): 295. Cape Town greatest video hits can be seen at https://www.youtube.com/watch?v=J3s5BAJpgFE.

CHAPTER EIGHTEEN: THE SECRET LIFE OF UNOWNED CATS

H. W. McGregor and colleagues' article describing using dogs to track feral cats was published in *Wildlife Research* in 2016, "Live-Capture of Feral Cats

Using Tracking Dogs and Darting, with Comparisons to Leg-Hold Trapping," *Wildlife Research* 43: 313–22; the results of their tracking studies appeared in 2015, "Feral Cats Are Better Killers on Open Habitats, Revealed by Animal-Borne Video," *PLoS One* 10: e0133915; and their paper on cats being able to move to recent fires was published in 2016, "Extraterritorial Hunting Expeditions to Intense Fire Scars by Feral Cats," *Scientific Reports* 6: 22559. McGregor put together a six-minute video montage to accompany the 2015 paper (https://www.youtube.com/watch?v=3KuypR5BBkU). J. A. Horn et al. "Home Range, Habitat Use, and Activity Patterns of Free-Roaming Domestic Cats," *Journal of Wildlife Management* 75 (2011): 1177–85, placed motion sensors on the collars of feral and pets cats in Illinois to calculate activity levels. The Jekyll Island results were published in S. M. Hernandez et al., "The Use of Point-of-View Cameras (Kittycams) to Quantify Predation by Colony Cats (*Felis catus*) on Wildlife," *Wildlife Research* 45 (2018): 357–65; and Hernandez et al., "Activity Patterns and Interspecific Interactions of Free-Roaming, Domestic Cats in Managed Trap-Neuter-Return Colonies," *Applied Animal Behaviour Science* 202 (2018): 63–68. Alexandra Newton McNeal provided details on the Jekyll Island project when I spoke with her on December 5, 2019, and in several emails before and after. McNeal is now an artist producing watercolors, oils, and acrylic paintings of coastal and marine wildlife (https://artbyalexandranicole.com/). She's quick to add that her days as a wildlife biologist are not definitely over and that who knows what the future will hold. National Geographic put together a three-minute video on the Jekyll Island project featuring McNeal, Hernandez, and the cats, https://www.youtube.com/watch?v=P8bd6dTcbd0.

CHAPTER NINETEEN: GOOD STEWARDSHIP, OR KEEPING YOUR RACE CAR IN THE GARAGE?

A. N. Rowan et al., "Cat Demographics & Impact on Wildlife in the USA, the UK, Australia and New Zealand: Facts and Values," *Journal of Applied Animal Ethics Research* 2 (2019): 7–37, documents that numbers of animals taken into shelters have declined greatly over the past several decades. S. M. L. Tan et al., "Uncontrolled Outdoor Access for Cats: An Assessment of Risks and Benefits," *Animals* 10, no. 2 (2020): 258, is an excellent review of research on the effects of keeping cats indoors. The title of this article says it all: K. Lauerman, "Cats Are Bird Killers. These Animal Experts Let Theirs Outside Anyway," *Washington Post*, September 2, 2016 (https://www.washingtonpost.com/news/animalia/wp/2016/09/02/cats-are-bird-killers-these-animal-experts-let-theirs-outside-anyway/). Studies on enrichment were reviewed in R. Foreman-Worsley and M. J. Farnworth, "A Systematic Review of Social and Environmental Factors and Their Implications for Indoor Cat Welfare," *Applied Animal Behaviour Science* 220 (2019): 104841. The study on different approaches to reduce cat hunting is M. Cecchetti et al., "Provision of High Meat Content Food and Object Play Reduce Predation of Wild Animals by Domestic Cats *Felis catus*," *Current Biology* 31 (2021): 1107–11. I talked to Martina Cecchetti by Zoom on February 18, 2021, with follow-up emails. Data on predation by cats in S. R. Loss et al. (2013), "The Impact of Free-Ranging Domestic Cats on Wildlife of the United States," *Nature Communications* 3 (2013): 1396; and S. Legge et al., "We Need to Worry About Bella and Charlie: The Impacts of Pet Cats on Australian Wildlife," *Wildlife Research* 47 (2020): 523–39. Two books on the debate about outdoor cats are P. P. Marra and C. Santella, *Cat Wars: The Devastating Consequences of a Cuddly Killer* (Princeton, NJ: Princeton University Press, 2016), and D. M. Wald and A. L. Peterson, *Cats and Conservationists: The Debate over Who Owns the Outdoors* (West Lafayette, IN: Purdue University Press, 2020).

CHAPTER TWENTY: THE FUTURE OF CATS

Among people who both suffer from asthma and are allergic to cats, the rate of emergency room visits for severe asthma attacks is almost twice that of those who live in a cat-free abode, according to P. J. Gergen et al., "Sensitization and Exposure to Pets: The Effect on Asthma Morbidity in the US Population," *Journal of Allergy and Clinical Immunology: In Practice* 6 (2018): 101–7. The study on behavior differences in kittens fathered by feral versus pet cats was summarized by John Bradshaw in *Cat Sense: How the New Feline Science Can Make You a Better Friend to Your Pet* (New York: Basic Books, 2013, pp. 274–75). An entrée to the large and entertaining literature on large black feral cats can be found in Darren Naish, "Williams' and Lang's Australian Big Cats: Do Pumas, Giant Feral Cats and Mystery Marsupials Stalk the Australian Outback?" (https://blogs.scientificamerican.com/tetrapod-zoology/williams-and-langs-australian-big-cats/), as well as in M. Williams and R. Lang, *Australian Big Cats: An Unnatural History of Panthers* (Sydney: Strange Nation Publishing, 2010); and K. Shuker, *Cats of Magic, Mythology, and Mystery* (Exeter, UK: CFZ Press, 2012). The "Lithgow cat" videos on YouTube (such as this one: https://www.youtube.com/watch?v=otXbm9Tn V0U) give a sense of the sorts of observations that have been made in Australia. A nice account of the fitoaty is available in Asia Murphy, "Makira Lessons: The Fitoaty (aka the Creature with Seven Livers)," Medium.com, February 14, 2017 (https://medium.com/@Asia_Murphy/makira-lessons-the-fitoaty-aka-the-creature-with-seven-livers-36da668baf2b). Arid Recovery is a fabulous research and conservation organization in Australia that has pioneered the use of cat-proof fences (https://aridrecovery.org.au/feral-proof-fence/).

Index

Italicized page numbers indicate material in tables or illustrations.